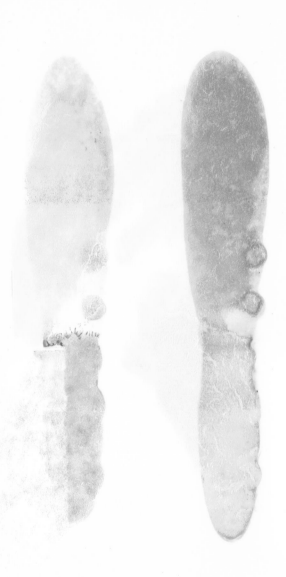

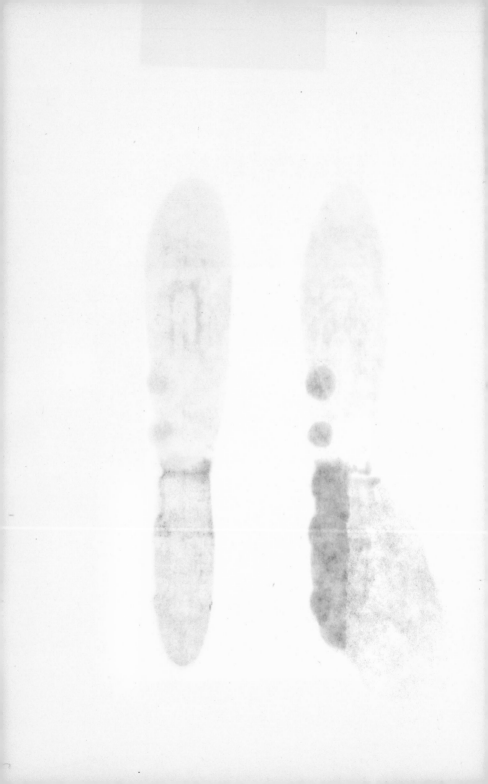

DOWNWIND

ALSO BY LOUISE MOERI

First the Egg
The Unicorn and the Plow
Save Queen of Sheba
The Girl Who Lived on the Ferris Wheel
A Horse for X. Y. Z.
How the Rabbit Stole the Moon
Star Mother's Youngest Child

DOWNWIND

by LOUISE MOERI

E. P. Dutton   New York

Library of Congress Cataloging in Publication Data

Moeri, Louise.
Downwind.

Summary: After fleeing their California home to escape
a possible radiation leak from a nuclear power plant,
twelve-year-old Ephraim and his family find themselves
caught up in circumstances perhaps even more threatening
to them.
[1. Atomic power plants—Accidents—Fiction
2. California—Fiction. 3. Survival—Fiction] I. Title.
PZ7.M7214Do 1984 [Fic] 83-20802
ISBN 0-525-44096-8

Published in the United States by E. P. Dutton, Inc.,
2 Park Avenue, New York, N.Y. 10016

Published simultaneously in Canada by
Fitzhenry & Whiteside Limited, Toronto

Editor: Ann Durell Designer: Isabel Warren-Lynch

Printed in the U.S.A. COBE First Edition
10 9 8 7 6 5 4 3 2 1

to my husband,
Edwin A. Moeri

CHAPTER 1

"I'm going to punch you in the eye!" shouted Ephraim Dearborn, even though he knew very well he couldn't do it. One punch in his brother Bones' face and Bones would knock him flat. What good was it, Ephraim wondered furiously, to be the oldest kid in the family—he was twelve—when your nine-year-old brother was as strong—well, *stronger*—than you were?

"Yeah! Hit him!" cried their sister, who was seven and plenty old enough to know better. She lunged against Bones but bounced off his solid bulk like a Ping-Pong ball off a board fence.

Bones, who was sitting squarely in front of the TV and had his hand clenched on the channel selector, glared at them over his shoulder. "I'm watching the *Noon News!*" he hollered. "Bug off!"

Ephraim gritted his teeth. Once again both he and Jocelyn had been outshouted and outshouldered by their brother Bones, who was almost as big as Ephraim and Jocelyn put together. The family didn't call the younger boy Bones—after the big, burly Brom Bones in "The Legend of Sleepy

Hollow"—for nothing; and his real name, Robert, had been all but forgotten around home.

Ephraim stomped into the kitchen. "Bones is hogging the TV again!" he told his mother. His blue eyes were wide and angry in his round face, and his straight fawn-colored hair looked as if someone had been yanking at it. His jeans and cotton shirt looked rumpled too.

His mother nervously brushed her short dark hair back. "Ephraim, you kids have got to hold the fighting down. I know you get bored during school vacation, but I just can't handle the uproar. Caleb's the only one of you who hasn't been yelling for the past hour, and the only reason he's quiet is because he's in his crib taking a nap."

His mother scooped up more peanut butter. The jar was almost empty, and she had to scrape hard to make four sandwiches. "We're going to have lunch in just a minute," she said as a blob of jelly slipped off her knife and dribbled down over her striped knit top and blue shorts. "Tell Bones and Jocelyn to come and eat."

"Well, that *might* get him away from the TV," grunted Ephraim.

Mom stacked the sandwiches on a plate and set them on the kitchen table, and then took a small pill bottle down from a high shelf in one of the cupboards. As she shook out one capsule and swallowed it, Ephraim frowned. Mom must be having a bad day if she had to take one of her "calm-down" pills this early.

Ephraim went back to the den. He reached for the TV knob to see if he could twist it fast enough to grab a quick glimpse of *Mr. Muscle,* but Bones' big fist was clamped on it like a vise. Furiously Ephraim glared at the TV screen. Some guy in a wrinkled white shirt and a tie half undone was reading an announcement. Ephraim, too preoccupied with Bones' tyranny to care, heard only a few words—"will keep you informed . . . situation at the Isla Conejo nuclear power plant . . . no danger—at this time—of a meltdown—"

"I want to see *Mr. Muscle!*" said Ephraim.

"Me, too!" cried Jocelyn. Jocelyn was small, slender, dark-haired and dark-eyed, and, like a cantankerous midget, she always sided with whichever brother had just attacked the other.

"Shut up! I want to hear the news!" Bones' voice boomed out. Everything about Bones was big—his broad-shouldered body in cut-off jeans and a dirty white T-shirt, his solid round head with springy, turflike brown hair, and most of all, his voice.

Ephraim aimed a kick at Bones, but Bones caught his ankle and Ephraim fell heavily against the TV, facing it. He screamed, not because he was hurt—he wasn't—but to bring Mom in from the kitchen. Plastered against the TV screen, with Bones and Jocelyn now fighting each other practically on top of him, he was the only one who could see or hear anything. Only one image was clear: a picture of two enormous cement towers, wide at the base, incurved, and then flaring out, located on a low grassy field. Then there were other pictures: the lobby outside the governor's office at the state capitol building, where people were milling around looking frightened; a street where a large number of police and fire personnel were hastily assembling equipment; and the rumpled newscaster again who said once more, "—wish to repeat at this time there is no danger—repeat, *no* danger —of a meltdown at Isla Conejo—"

"All right, you kids—knock that off! Get in here and eat lunch, and I don't want to hear another word out of any of you!" bawled their mother, stomping in from the kitchen.

Ephraim felt Bones and Jocelyn roll away from him, and he peeled himself off the TV. At that moment Mom punched the Off button and the screen went dead. As he got to his feet, glaring at Bones, he started to yell but stopped when Mom's knuckles rapped his head. I'll get him later, Ephraim said to himself. Just wait—

In the ktichen they settled into their places, and their mother sat down too. Ephraim knew she didn't like to eat with them but did it to reduce the chances of a fight breaking out. Quarreling made her nervous, but then so did everything else. Ephraim could have run a Boston marathon with the number of miles he had racked up just this summer checking up on Bones and Jocelyn and Caleb for Mom.

Being the oldest kid in the family was a drag. He got to stay up later at night and had a bigger allowance; but balanced against those small privileges, he had to "set a good example" for the three younger ones, and, above all, Mom always expected him to know where they were and what they were doing all the time. Sometimes he thought about suggesting she fasten collars around their necks that would have those little radio transmitters in them—like the scientists put on grizzly bears to track them. Then he could have monitored the kids on a radar screen and saved a lot of walking. Of course he knew better than to actually say a thing like that. Mom would really blow up. But he was sure tired of being a baby-sitter. One of his most closely guarded secrets was that he knew how to change diapers. He hated to think what the other guys in the sixth grade would say if they found *that* out!

Well, maybe someday things would change, and he would get to take Red Cross swimming lessons or go deer hunting with Dad, but for now he was shut inside the backyard fence with Bones and Jocelyn and Caleb. He often wondered what the world was like outside that six-foot board fence.

If Mom had her way, he'd never find out. Some kids' mothers gambled the grocery money away on weekends at Las Vegas, and some drank. His mom was a worrier. She was always finding things for them to do, because it made her feel safer for them to be busy but close, where she could see them. That was one reason why they had such a neat house, flowers blooming in the yard, no weeds.

"After lunch," said Mom now, "we're going out to do some chores in the backyard."

"Good," said Ephraim. "No sense watching TV— Bones hogs it all the time."

"Pig!" shouted Jocelyn, pointing at Bones.

"Am not!" cried Bones. "You guys just want to see that crazy *Mr. Muscle. I* was watching the news—"

"Like you know so much about the news!" snorted Ephraim.

"Dummy!" cried Jocelyn.

Bones kicked at her under the table. "Dad says to watch the *Noon News* 'cause he's going to ask us questions about it when he gets home!"

This was true, and usually Ephraim went along with the plan, but today he was just plain tired of the way Bones pushed everybody around. Maybe, he thought, if he watched *Mr. Muscle* enough, he could do some pushing of his own.

"So? What did you see on the news, then?" asked Ephraim.

"Yeah, what?" Jocelyn choked out around a mouthful of peanut butter and bread.

"About . . . uh . . . some guy said . . . bulletins . . . uh . . . will come later . . . uh . . . uh . . ."

5

Ephraim looked at him with contempt. "See? You don't even remember what he said!"

Bones straightened up. "Well, *you* were between me and the TV!"

"You *pushed* me!" shouted Ephraim.

"Stop it! Stop it!" cried Mom. Her face was getting red and her hands were starting to shake. They looked at her and stopped. "Okay," said their mother carefully, "Ephraim, since you were the closest"—she choked a little—"closest to the TV, what did the newscaster say?"

Ephraim struggled to pull the words together quickly. He would only have a few seconds before Bones would grab the center of attention again. What was that word? Oh, yes— *"Meltdown.* That's what he said. *Meltdown* at the Isla Conejo nuclear power plant." He looked up at his mother, expecting to see at least a pleased smile. Instead, her face started to freeze. Her eyes widened, and her mouth, though silent, turned into a scream.

"Mom?" whispered Ephraim. "Is something wrong? Mom? *Mom*—?"

CHAPTER 2

Ephraim watched silently as his mother closed her mouth, and her eyes took on a hiding look behind her lowered lids. Then she rose, sandwich in hand, though he could see she had forgotten it, and went quietly into the den. After a moment Ephraim followed. Bones and Jocelyn were starting another fight and appeared not to notice them.

In the den, his mother turned the TV back on. It was still tuned to the same channel, and the same man in the white shirt, his tie still dangling crookedly, was reading again from a sheaf of papers. "We will, of course, be in constant communication with the public information office at Isla Conejo and will continue to issue bulletins as soon as they are released. At the present time, the National Guard has not been called out, although as a routine safety measure, all off-duty police and fire-fighting personnel have been asked to report for duty. There is no—repeat, *no*—imminent danger of a meltdown."

His mother reached out, laid her hand on the top of the TV, and did not even notice that her fingers were on a sticky, wilted banana skin that Bones had left there. Jocelyn's dolls, at her feet, went unnoticed too.

Ephraim, reading her as well as listening to the words from the TV, spoke. "Mom? What's a *meltdown*?"

His mother drew in a sudden deep breath. She had not noticed that Ephraim was behind her. "Oh—it's—uh—a kind of—accident."

"Accident? You mean like an explosion?"

"Well—not exactly. But he says"—she was holding herself in tightly, and Ephraim suddenly felt that they must be in the gravest danger—"he says there's no danger of one happening."

She had omitted a word from the TV newscaster's speech and Ephraim caught it. "He said, 'no im—im—minent danger,' Mom. What's *im*—?"

"Oh . . . *imminent* . . . that means, well, 'right now,' going to happen . . . *immediately.*"

She put just a little too much emphasis on the last word and Ephraim pieced it together. "Then what he means is that it *could* happen later?"

His mother stared at him wordlessly, and Ephraim knew she would not answer him. Sometimes Mom got so upset that she seemed to hide inside herself, unable even to tell them what had frightened her.

Ephraim chewed his lip for a moment. He wanted to know more—especially that part about calling out the National Guard—but decided not to press his mother just yet. He could see she was going to stay by the TV to see what happened next, and a good idea struck him. He turned and galloped back to the kitchen. As he grabbed his sandwich he said to Bones and Jocelyn, "Me and Mom are going to watch the news."

Bones and Jocelyn glanced up and then went on fighting. They weren't really interested in the news, and would keep themselves busy arguing and then probably go outside.

Back in the den, Ephraim pulled up a stool for his mother and a TV cushion for himself. His mother's knees buckled and she sat down on the stool, still holding her half-eaten sandwich. Then the rumpled TV newscaster disappeared and a commercial for floor wax came on, followed by another commercial for a new brand of dog biscuits. During the commercials, his mother began to eat her sandwich slowly and deliberately, like somebody taking medicine, and Ephraim started to eat his.

In just a few minutes, the string of commercials ended and another newscaster came on—he was staring down at the monitor. After two or three seconds of silence, he raised his eyes and said quietly, "This news broadcast will resume after these commercial messages."

The newsroom faded out and was replaced by a commercial in which a middle-aged lady was dancing with her cat. Mom reached out and turned the sound down to a whisper, although she continued to watch the screen. Ephraim knew she would turn the sound up again when the newscast resumed.

"Mom, you . . . do you think . . . Why did they call in the policemen and firemen?" Ephraim had finished his sandwich and carefully licked the peanut butter off his fingers. His mother was still mechanically chewing and swallowing bite after bite of hers.

"I'm . . . not sure."

"But Isla Conejo is up toward Sacramento, isn't it? Not near here?"

"Yes, but—"

"Then . . . the trouble . . . is in Sacramento?"

His mother was silent for several seconds, staring at the TV. Then suddenly she turned to him as if she had come to some sort of decision. "Ephraim," she said abruptly, and her voice was quite low, as if she were forcing it out. "Ephraim, there is—something we—both of us—have to do. It's important. No, don't ask too much now. I don't really know much for certain yet. But—just in case—I'm going to make some phone calls. Try to find out all I can. You go out now and make absolutely *certain* that Bones and Jocelyn stay in the yard. Threaten them. Hit them. Anything. Just make sure they stay right here at home."

Ephraim's eyes widened. His mother was telling him to *hit* his brother and sister, if need be, to make them obey!

"Yes," went on his mother, seeing his astonishment. "I'll keep on listening to the TV, and I'm going to make some phone calls."

"Mom—what—?"

"I'll tell you anything I can—as soon as—as soon as I find out what is going on."

Ephraim had a million questions that were choking him, but he didn't get even one out before she grabbed him by the shoulder and pushed him out of the den. From the look on her face, Ephraim knew he had better keep Bones and Jocelyn in the backyard, or all hell was going to break loose.

CHAPTER 3

Under the walnut tree was the shadiest place in the backyard. There were other trees—almonds and peaches—but the walnut's heavy, leathery leaves made the ground beneath it the only really cool place.

The children were spending a lot of time under the walnut tree this summer because for some reason the tree had been adopted as a favorite resting place by a flock of sparrows. All day long the sparrows came and went among the gnarled, knobby branches, and their soft, shimmery voices made Ephraim think the twigs and leaves had voices and were spending the summer telling each other stories.

Now, as the sparrows showered down little crisp phrases, the three children were lying flat on the grass, staring up into the canopy of dense green. A few feet away, their dog, a small black mongrel named Turk—short for Turkey—was sprawled in a hollow he had dug in the moist dirt under the spirea bushes. The bushes grew along the back side of the house, and at the end of their bed was the doghouse, and beyond that the patio opening off the kitchen. From the open kitchen door leading out onto the patio, they could hear the blurred sound of the TV, and then from time to time their mother's voice as she dialed first one number and then another on the phone. On each call she spoke briefly; a couple of times she slammed the receiver down. Twice the phone rang and she rushed to answer it, spoke in a guarded voice, hung up.

Bones and Jocelyn had no idea what was going on, and Ephraim was glad they didn't. They would be sure to ask questions of him and he needed time to figure it out—as much as he could—first. Lying there with the spangles of sun and blue sky flickering above him, Ephraim tried to recall bits of information from school science classes, movies he had seen, science fiction books and comics, conversations between his parents and their friends.

Pieced together it looked like this: thirty miles north of them was a place called Isla Conejo—a Spanish name for rabbit island—and at Isla Conejo was a nuclear power plant. He had seen it often when the family went up to see friends and relatives in Marysville, north of Sacramento. The plant was a cluster of low buildings behind a perimeter of cyclone fence, and in the center of the cluster rose two huge cement towers. He never saw anything more because the plant, the towers were located at least a mile from the highway.

He knew very little about how nuclear power plants worked. A special kind of fuel was burned in them, and this produced electricity. He had planned last year to do a model of a nuclear power plant for the Science Fair, but Matthew Frost picked it first, so he had done photosynthesis instead. He tried to recall Matthew's diagrams, but all he could remember was a domed, circular building called a containment building, in which was something called a reactor. Most of Matthew's display seemed to be diagrams of pipes leading to a system of steam turbines that generated electricity, and there was something about a cooling system based on pressurized water, whatever that was.

But he did remember some other things, closer to home, about the electricity Isla Conejo produced. His mother and father and their friends frequently complained that building Isla Conejo was supposed to have provided cheap electricity. Instead, everybody's electricity bills had gotten higher and higher. Mom had even started hanging the laundry on what she called a solar dryer—a clothesline—and always tried to bake several things in the oven at one time. Dad was talking about cutting back on using things like the air conditioner, the TV, and even the outside lights they kept on at night for safety. They had never done without these lights in the past, even though out here in the country they felt safe.

Or at least they had felt safe until today. Ephraim did not recognize the word *meltdown,* but he did recognize fear when he saw it. Those people in the TV newsroom were afraid—talking about things like calling out the National Guard. Mom was afraid. In itself that was nothing new, but the people she had talked to on the phone must have been afraid too, judging by what Mom had said and the tone of her voice.

Therefore, Ephraim was afraid too.

He sat up abruptly. "Bones," he said, forgetting how mad he had been a little earlier, "I'm going to try and hear what Mom's talking about."

"Sneak!" said Bones instantly.

"I'll tell!" said Jocelyn.

"Listen," said Ephraim, "there's something funny going on."

"You ain't supposed to listen in when people are talking. It's called roof-dropping!" said Jocelyn.

"You mean eavesdropping," said Ephraim impatiently,

"and I'm going to do it anyway. There's something going on."

"Okay. If you can do it, we can too." Bones rose up and faced his brother truculently. As Bones and Jocelyn followed Ephraim onto the patio, Ephraim comforted himself that at least he was keeping them here at home, like Mom said. He knew that if he had dropped the least hint that he had been told to do that, Bones and Jocelyn would have run off like stampeded cattle.

On the patio they crouched into a tight little cluster, with legs, arms, knees and chins knocking each other here and there. But they could hear very well.

"What do you mean, 'there's *no* evacuation plan'! We've got to get out of here, all of us—head for the hills— *What?*" There was a tense moment as she halted, seemed to be listening, and nodded once or twice. "Yes—we're *all* downwind from Isla Conejo—the whole valley will be blanketed with radiation if that place blows up! We'll all be killed! We're going to take our family and *run*—head for the hills! Now!"

Their mother slammed the phone down and then bolted off to some place inside the house where they could not see her or hear her. Ephraim thought she was probably checking on Caleb, who would be waking up from his nap soon.

The three children untangled themselves and stared at each other.

"What's she mean, 'head for the hills'?" asked Jocelyn, puzzled. "Dad's vacation is *next* week—"

"Hills?" said Bones, frowning. "Today is Tuesday—not even close to Saturday. We never go to the hills on Tuesday."

But Ephraim got to his feet slowly, though the panicky feeling he had been fighting down was getting stronger. "She said, 'head for the hills,' " he said quietly. "And I think she

14

means what she says. Come on—if we're going to the hills, we'll sure take the trailer. And we can get a lot of stuff loaded into it now. Then when Dad comes home, we'll be ready to go."

He rose to his feet and lit out as fast as he could go toward the far end of the lawn, yanked open the big gate, and plowed on through. The trailer, a seventeen-foot, self-contained Wilderness, was just ahead, tucked in between two peach trees, where Dad always parked it.

Behind him Bones and Jocelyn were coming, but both of them were much slower. When he looked back, Ephraim saw why. Jocelyn had dashed into the house and grabbed the cockatiel's cage and his can of birdseed, and Bones had their gray cat, Kalijah, wildly struggling under one arm while he dragged the protesting Turk along with the other.

"Mom says we'll all be killed! So if we got to go, *they* got to go!" panted Bones.

Ephraim started to yell but decided not to. He could see that Bones had realized there was something dangerous happening, and he would never leave the animals behind if he thought they would be hurt. Bones was scrappy and tough with people, but he was funny about animals. It was all they could do to keep him from bringing home all the stray cats and dogs he found. Bones wouldn't even swat flies—he chased them out of the house with the broom.

But never mind Bones— Ephraim had other things on his mind. He knew the trailer was always partly ready to go, since Mom had put all her old pans and dishes in it, together with a few staples like sugar and flour and coffee. The shovel, bucket and axe required for fire safety were always in the compartment under one of the seats, but he would check to see what else was needed.

He had already learned how to pick the lock, and it took only a few seconds to fling open the trailer door. As Bones pushed the anguished cat into the closet and shut the door and tied Turk to the table leg with his leash, Jocelyn hauled herself and the birdcage over the high threshold. The bird was thrashing around, scattering birdseed and small white feathers and yelling, "Corkey's a dirty bird! Corkey's a dirty bird! Hello, wino! Hello, wino!"

"Oh, shut up!" screamed Ephraim nervously. "Bones, check the canned stuff! Jocelyn, make sure the blankets are still here from the last trip! I'll fill the water tank and get the —some other stuff—" He didn't want to tell them exactly what it was he planned to get, but the words *National Guard* were still echoing in his mind. Things could get pretty bad, he felt, if it came to that.

But as he raced toward the faucet to hook up a garden hose to fill the water tank, he knew one thing for sure—they would be ready for *anything* when it came time to "head for the hills."

CHAPTER 4

He had just turned the water off and was pulling the hose back to coil it up when he heard a peculiar sound. It was a man's voice, shouting, threatening, pleading, somewhere out behind their house.

Ephraim dropped the hose and galloped back to the garden, opened the gate and slid through. This, the vegetable garden, was at the very end of their lot, which was bounded on the east and south by an immense walnut orchard. The orchard belonged to an old man by the name of Bonfiglio. Next to God and his dad, Ephraim was most careful not to antagonize Mr. Bonfiglio. Mr. Bonfiglio had been known to chase trespassers out of his orchard with a shovel raised, threatening to bash in their heads. Ephraim had never set foot in the orchard, and he planned to keep it that way.

Mr. Bonfiglio was at least seventy years old, maybe eighty, and he had spent his whole life caring for this orchard. He had come from Italy as a boy, so the story went, got a job irrigating these trees when they were saplings like himself, and had starved and slaved and suffered to buy the orchard for his own. As the years passed, the trees grew into giants and Mr. Bonfiglio grew into a gnome—wrinkled, bent, brown and frowning. He must be very rich—an acre of walnut trees, sold for cash, Dad said once, would have paid off the Dearborns' mortgage and put a new roof on the house.

But being rich didn't show on Mr. Bonfiglio. He wore the oldest clothes and drove a Toyota pickup that had been wrecked twice. He did his own irrigating and cultivating, and lately had bought a flock of geese which, working for free, so to speak, was supposed to eat the weeds that sprouted up after every flooding. Sometimes the old man let the geese paddle around when the irrigation water was deep between the trees, and Ephraim had liked watching them, buoyant and graceful as sailboats in the deep shade.

Now Ephraim realized that what he had heard was Mr. Bonfiglio calling his geese. Ephraim climbed up two levels of wire and leaned over the top of the fence so he could see better. Yes, there he was. The old man was running wildly here and there under the giant trees, calling and shouting to the geese. His ragged shirt flapped behind him and his nearly white hair straggled down over his face.

Then Ephraim saw the birds. There were one gander and five geese, and Mr. Bonfiglio had evidently let them into the orchard that morning to eat the new weeds. Now he was trying to round them up and herd them (can you *herd* geese?) back to their pen earlier than usual, and the geese were determined to stay in the orchard.

Mr. Bonfiglio charged at the nearest goose, a graceful matron with smooth gray feathers and proud black eyes, and the gander, outraged, charged at Mr. Bonfiglio. The gander's wings were raised to strike, and his beak was open as he hissed a terrible challenge. Mr. Bonfiglio faltered. The ruffled goose twitched her tail feathers and bent her head to snap up a juicy weed.

Mr. Bonfiglio galloped around the little flock and, arms wide, tried to approach from the opposite direction. The gander whirled and raced at Mr. Bonfiglio. The male bird was so big that his beak, powerful as a hammer made of bone, was on a level with the bent old man's face. The old man retreated.

Ephraim knew better than to climb over the fence and go to the old man's aid. Mr. Bonfiglio never allowed any but hired hands—and them only as needed—to set foot in his orchard.

Then Ephraim saw something that might help. Dad had left a tree stake lying on the ground on their side of the fence. It was about six feet long, slender, a good weapon to use against the wrathful gander.

"Here!" cried Ephraim. He grabbed the stake and threw it like a javelin over the fence. It arced down and dug into the plowed ground at Mr. Bonfiglio's feet.

The old man grabbed the pole. He lunged forward and poked the gander at the spot where his magnificent S-curved neck joined his body. The gander let out a squawk and dove at the old man.

Mr. Bonfiglio tried to retreat, tripped over his own feet, and fell backwards in a heap of old clothes with flailing arms and legs. The gander's deadly beak was aimed straight at the old man's eyes.

Instantly Ephraim was over the fence and running. But instead of going toward the old man, in a flash he conjured up a better plan. Shooting past the gander, he grabbed one of the geese, tucked her head backwards under his right arm, and lit out for the wire goose pen that faced the orchard.

The outraged goose let out a cry that rose and fell like a fire siren. The gander left off his attack on the old man, whirled around, and saw one of his flock, head dangling and feet paddling in midair, being kidnapped. He started after Ephraim. If the gander caught him, Ephraim knew he would be black and blue for weeks.

But with several seconds' head start, Ephraim reached the pen, shinnied up the wire, and flung the goose over the fence. She landed butt down, squawking and shrieking and then scrambled onto her yellow platform feet and rushed back to the fence.

Ephraim, perched high up out of the gander's reach, felt a hot breeze fan his face as the gander leaped and snapped his beak and beat his wings only a few inches below him.

Ephraim, teetering at the top of the fence, made a few gestures just to keep the gander busy. Out of the corner of his eye, he could see Mr. Bonfiglio climb to his feet and stagger around, gathering the other four geese with the help of Ephraim's stick. In moments the old man had the other four geese locked in with the first. The gander, upon seeing all his ladies now inside the fence, reluctantly abandoned his plan to murder the old man. Hissing like a steam engine, he backed into the pen. Mr. Bonfiglio slammed the gate.

Silence. Ephraim, perched at the top of the fence, looked down. On one side was the angry gander, wings out and beak ready to strike.

On the other side of the fence stood Mr. Bonfiglio, who guarded his orchard the way the gander guarded his geese.

Ephraim wondered bleakly which side of the fence he ought to climb down.

Then, slowly, Mr. Bonfiglio shuffled forward. He peered up at Ephraim. He held out a hand that was as brown and hard and gnarled as the root of one of his trees, and Ephraim couldn't guess if the hand was going to whack him on the rump or grab a handful of his hair—for trespassing.

Then Mr. Bonfiglio spoke. "The—radio—he said—"

Ephraim frowned. The old man's voice was rough, his accent heavy. And that clublike hand—?

"—radio—he said—big trouble!"

"Oh. Yeah. At Isla Conejo. We heard it too," said Ephraim. Behind him the gander had launched a new attack by running at the fence. Ephraim wobbled as the fence swayed.

"Had—to get geese—in—pen—*early*. My daughter says Mama and me—got to go—away with her!"

Ephraim did not know that Mr. Bonfiglio had a daughter but now as he glanced around, following the old man's look, he saw a plump pretty woman, older than Mom or Dad, hurrying down the back steps of Mr. Bonfiglio's house. She beckoned urgently to the old man to come.

"Come on, Papa! We got to go! Mama's ready—I'm taking you to Tony's house in Truckee—"

Mr. Bonfiglio nodded. "All right—all right! Come down, boy—it's okay—Garibaldi, he penned up. He can't hurt you now."

"Garibaldi?" Ephraim stared.

"Ol' Garibaldi, he ain't mean—he just tries to take care of his own."

Ephraim turned for a last look over his shoulder, where Garibaldi was still charging the fence. The fence posts were beginning to lean.

"Come down, boy." Mr. Bonfiglio reached out again, and this time the crooked old hand settled gently on Ephraim's arm, and Ephraim found himself climbing down on Mr. Bonfiglio's side of the fence.

21

"You—go? Your Mama and Papa take you—away—because the radio, he say—?"

Ephraim nodded. "Yeah. Mom says we're going to the hills as soon as my Dad comes. Mr. Bonfiglio—I'm sorry I came over into your orchard. I know I wasn't supposed to. But I was afraid—that gander—Garibaldi—was going for your eyes!"

Mr. Bonfiglio sighed and his hand, raised from Ephraim's arm, lightly brushed his hair. "No—no—you good boy. Helped me—you good neighbor—"

"Papa! We've got to go!" the lady said.

Mr. Bonfiglio turned, pushed Ephraim gently toward his own fence. "Yes—you go too!" he said to Ephraim. "Tell your Mama I said—you a good boy."

"Papa!"

Then the old man turned to follow his daughter, and Ephraim, released, ran toward the garden fence. As he climbed over and dropped down on his own side, he looked back once.

Garibaldi was still patrolling his perimeter, and Mr. Bonfiglio had disappeared.

"Someday," Ephraim said to himself, "I'll talk to him again. Mr. Bonfiglio's not so bad. And I'd never have found it out if I hadn't had to help him get Garibaldi into his pen. I might even get to like Garibaldi."

CHAPTER **5**

Ephraim loved his mother, although he wasn't much like her. He had Dad's sandy hair and blue eyes and quiet ways. Bones took after Mom—in personality, at least—and they were both like pistols that didn't shoot straight: they exploded first and aimed later. Sometimes they hit the target. Sometimes they missed. Bones had already dropped a can of beans on Turk's tail and skinned his knee on the gate, but Ephraim could see that he had succeeded in getting a lot of food into the trailer.

When Ephraim was satisfied that they had done all they could to get the trailer ready, he led Bones and Jocelyn back into the house. He figured that Mom would be getting ready to go, but instead she was disorganized and nearly hysterical.

She had started to pack suitcases in the living room, but they were only half full. Bread, a carton of eggs, some cheese, a half gallon of milk, were scattered over the kitchen table and chairs, and now she was in the bathroom yanking bottles and jars out of the medicine cabinet and throwing them into paper bags.

Caleb had awakened and was standing up in his crib yelling for Mom. Caleb was the only one of them to have curly hair, and as Ephraim glanced through the bedroom door at his baby brother, he thought Caleb's hair looked like the swirly wood shavings Dad made sometimes when he was smoothing a board. Caleb's face was flushed and he was only half awake, but he could still make the walls rattle with his bellowing. Ephraim thought about getting him out of his crib but decided to leave him where he was. Otherwise he would quickly unpack everything Mom had packed.

"Good boy!" he called to Caleb.

Caleb yelled.

At the bathroom door, Ephraim reached out to grab his mother's arm. "Mom?"

She started. When she turned, he could see she had been crying. "Ephraim—what—where are Bones and Jocelyn?"

"In the house. What are—do you want me to help?"

"No! Yes! Oh, God, I don't know!"

Ephraim reached out and scooped up two unopened bars of soap and dumped them into her bag. If I'm old enough to watch the little kids, he wanted to tell her, I'm old enough to know what's going on. But he didn't think she would listen, or if she did, she'd just get mad.

His mother raised her hands and wiped the sweat from her face. She took a deep breath and tried to steady herself. Then, speaking slowly and carefully, she said, "There has been an accident—a bad accident—at Isla Conejo. We are downwind from Isla Conejo, and that means if radiation escapes—like radioactive steam or gas—it will blow toward us. Everybody in the Central Valley is in great danger—"

"How . . . great?"

24

"Very . . . great."

"Oh." Ephraim thought about telling her how he had just talked to Mr. Bonfiglio, and that they were leaving too, but decided against it. Better stick to what was going on right here at home. "Mom—shouldn't we call Dad?"

"I already have. The factory was closed at ten minutes to one and everybody sent home. He'll be here pretty soon."

"Mom, I heard you say we were going to leave—"

"We've got to! We've got to!" Her voice was creeping up into a shrill whisper.

"Okay, okay," said Ephraim, hoping to dampen down her panic. He had already lived through many of Mom's outbursts, and had adopted his father's method of dealing with them: don't yell, don't argue. Go along with her, try to calm her down. And Dad would be here pretty soon.

"Listen, Mom, the trailer is ready. Us kids went out and checked it over, and loaded some stuff."

"Trailer?"

"Yeah. You said we're going to head for the hills. So we got the trailer ready. Water, canned stuff. Blankets. Do you want us to get the rest of this stuff loaded into it?"

His mother flashed a look of gratitude. "Oh, the trailer! Of course—that's what we'll do! Yes—get everything loaded into it, and then all Dad will have to do is just hitch it to the pickup when he gets here!"

Ephraim took his mother's paper bags, went back to the living room and gathered up a pile of sweaters. As he headed back out to the trailer, he nudged Bones toward the den. "Keep an eye on the TV, in case they give out any more stuff about Isla Conejo," he said. "I'll help Mom—she's about to blow her stack. Jocelyn, you watch for Dad—"

"Daddy's coming home? Now?"

"Yeah. Right away. And we want to be ready—"

"Ready for what?" asked Bones as he headed for the den.

Wish I knew, said Ephraim to himself. Wish I knew.

CHAPTER 6

"He should have been here by now." Their mother's voice was tense as she stood at the front window staring down their country road. "They closed the factory half an hour ago. Why isn't your father home yet?"

She had gotten Caleb out of his crib and changed him and put his Oshkosh overalls on him, and his shoes and socks. Caleb was used to going barefoot most of the time and was now having a tantrum, bawling and kicking on the floor as he tried to figure out how to get his shoes off. The laces were tied in double knots, so he couldn't.

Ephraim, standing next to his mother, could just see the intersection about one hundred yards away at the spot where their road crossed the state highway, and he had a pretty good idea why Dad had not yet arrived, but he didn't think it would do any good to say so to Mom. Through the tops of the grapevines between them and the highway, he could see an unbelievable flow of traffic: ordinary cars, buses, emergency vehicles, bikes, trucks and pickups were surging both ways, as if terribly urgent business had broken out like a grass fire and people had to rush off in all directions to deal

27

with it. Dad was a good driver, but it would take him a long time to thread through that traffic jam.

Mom stopped her watching from time to time to rush off and put something else into one of the suitcases. She had packed cartons of clothes and diapers and food for the baby and had folded and stacked the Port-A-Crib beside the cartons. Ephraim knew there was room for all the stuff, between Dad's pickup and the trailer, but seeing it there made him wince. Dad and Mom were sure to have a fight about it. Dad always said she wouldn't go to the grocery store without taking half the house with her.

"Hey, Mom, where are Turk's dog biscuits?" asked Bones, coming in from the porch with the dog's food and water bowls under one arm. Evidently he had left off monitoring the TV for a moment to make sure that Turk's needs were met in spite of whatever disaster was about to occur.

"Why?" muttered their mother distractedly.

"So I can put them in the trailer," said Bones. "And Kalijah's cat food."

Mom looked blank and then let out a yell. "Dog! Cat! This is no time to worry about a dog and cat! Forget them—we can't take them."

Bones and Jocelyn let out a howl at the same time. Mom started to yell back at them when a noise outside distracted her. She rushed to the window to see if it was Dad.

During the momentary lull, Ephraim slipped out to the laundry room and located the sack of dog biscuits and several cans of cat food. He threw the cat food cans into the dog biscuit sack, gave it all to Bones, and pushed him out the side door. "Load it into the trailer," he hissed, "and keep your mouth shut. If you can keep Turk and Kalijah out of sight —in the closet—Mom and Dad won't know we're taking them!"

28

Back in the living room, Ephraim saw his mother make a lunge for the door. She yanked it open, and Ephraim's father stood on the threshold.

"Hi," said Dad quietly.

Fifteen minutes later, their mother was charging up and down the living room, yelling and waving her hands. Bones and Jocelyn sat on the floor by the window, and Ephraim leaned over the back of the couch. Dad, who had insisted on pulling off his work coveralls and changing to his everyday jeans and cowboy boots, had sent Ephraim to bring him a can of beer, while Caleb crawled up on his lap and began to gabble about his shoes. Now their father, alert but not jangled, sat watching his wife, as Caleb forgot his shoes and started to go through his father's pockets.

"Gabriel! We've got to go! Now! Head for the hills! Haven't you heard? Don't you know what it means? I've been telling you all along that those nuclear power plants aren't safe—things go wrong with them all the time! I've studied magazine articles—gone to meetings—those buildings are falling apart—the machinery isn't safe! But the people in charge keep covering up the facts. I've tried to tell you about it, but you just don't listen! You won't go to the meetings with me, and you always find something else to do when somebody explains it on TV. Just last week—"

"Well, how do we know who to believe, Carrie?" asked Dad sharply. "You hear one side—the plants are safe as a mother's love. You hear the other side—we're all going to hell on a poker. Who's right?"

"Listen, Gabe," said Mom, and her voice was shaking, "look out the window. Listen to the radio, the TV, and decide who's right about it today. In Sacramento they've called up the reserve fire fighters and police—they may call out the National Guard. They're already talking about a meltdown! You know what that means—there's no way to stop it once it gets going! And the buildings may explode— may have already— My God, do you want your children to die?"

Dad's voice was level. "I listened to the radio coming home. But they haven't said it was a meltdown yet."

"Oh, Gabe, for God's sake, when they *deny* something like that it means everybody knows a meltdown is possible!"

Dad sat silent for a moment. Ephraim figured that he knew she was right about the possible danger, and also that the people who were giving out information might not be telling the whole story. Mom had said that for a long time. But Dad was usually quieter than Mom, and more inclined to dig in rather than run. Mom, Ephraim well knew, always saw disaster and ruin, whether the situation was really that bad or not.

Then in quick succession Dad listed the authorities he always said they should consult. "Did you call the sheriff's office? The fire station? Police department in town? Civil defense?"

Mom turned on him. "The sheriff's office has had only busy signals since it happened. Same for fire. The police say listen for information on the radio. And there isn't any civil defense."

Gabe stared at her. "What do you mean there isn't any civil defense? There has to be!"

"No! There isn't! Nothing. The civil defense department was closed down a half-dozen tax cuts ago."

"But if we have to—evacuate—what is the plan? You can't have a big mass of people driving around in a panic—it's dangerous. There have to be routes, rules, leaders, destinations—"

Mom faced him desperately. "Gabe—that's what I'm trying to tell you! There is *no* civil defense department. No routes. No plans. Gabe—this may be the last day of the world—and we're completely on our own!"

CHAPTER 7

Dad put the receiver back onto its cradle and let out a short, carefully smothered snort of held-in air. He was not frightened, Ephraim could see that, but he was concerned about the responses to his phone calls.

He stood silent a moment, smoothing his hand over Caleb's curly head. Then he said, "The Millers are leaving. They have family in San Diego and they're going down there. Bill Woods is taking his family—doesn't know where. George Palmer says he and his wife are going to wait it out. They have a big, deep cellar and he thinks they'll be safe."

"Oh, God—"

"I can't reach Duane Haskins or Ray Fuller. Nobody answers."

"Did you try the sheriff again?"

"Busy."

"Police?"

"Said listen to the radio."

"We've had it on all the time. They don't *say* anything!" Mom's voice was hoarse. It sounded like she had a bad cold, but it always got that way when she was nervous. She turned and reached across the dining-room table to turn the volume up on the portable radio.

"—Watson, of the Nuclear Regulatory Commission," said a man's voice on the radio, "has just arrived by chartered plane at Executive Airport in Sacramento. Mr. Watson, who has been driven directly to Isla Conejo, could not be reached for any comment by our reporter while at the airport, but we were assured by members of Mr. Watson's staff that as soon as he has had a chance to fully assess the situation, he will meet with all press representatives. In the meantime, we will continue with the usual programming—one moment, please. I have been asked to read the following announcement: Will Dr. David Stern please report to Sierra Hospital? Dr. David Stern, Dr. David Stern, please go immediately to Sierra Hospital. And now, ladies and gentlemen, let's listen to Merle Haggard sing—"

Mom frowned. "David Stern? That name is familiar—"

"TV," said Dad briefly. "Couple of nights ago."

She stared at him. "Burn unit," she said. "David Stern heads the new burn unit at Sierra Hospital. Gabe—let's go!"

Ephraim had stood in silence, listening to every word Mom and Dad said as if he would be asked questions about it later. And he would. Bones and Jocelyn never listened to either Mom or Dad if they could help it, except for when they were getting direct orders. Ephraim knew they would both depend on him to filter out some information and keep them posted.

Oddly enough, Dad and Mom used the same system only in reverse. Both of them—Mom, especially—seemed to expect Ephraim to know what Bones and Jocelyn were doing, and report back as needed. They were equally likely to issue instructions regarding Bones and Jocelyn through Ephraim, and hold him responsible for what happened. Ephraim didn't care much for his role as go-between, but so far he hadn't been able to break out of it. Now he watched as Bones and Jocelyn sidled out the back door and headed for the trailer. They were going to check on the animals, he figured. Ephraim kept them in his line of sight to be sure they didn't wander off. He had no idea how long it was going to take for Dad and Mom to decide what to do.

The phone rang.

Dad, standing closest, grabbed it up. "Hello? Mother? Yes —we know—we're—trying to get some instructions—you heard what?"

Mom watched him tensely. Ephraim knew Dad was talking to Grandma Dearborn, who lived in Marysville, north of Sacramento. He wondered if she was in danger too, and started to ask his mother, but she guessed what he was going to say, and answered first. "No, she's all right. They're way north of Isla Conejo and the wind won't be blowing toward them. We're downwind—we're the ones in danger—"

Ephraim turned to listen to Dad, who wasn't saying much but, "Yeah. Uh-huh. Yeah. Oh? Yeah—" In a few moments he hung up.

"We got disconnected," he said.

"What did she say?" asked Mom.

"She said somebody was interviewed on their local station and he said the whole center of the state should be evacuated."

"That's *us*, Gabe! Listen, Gabe, the children—do you want them to be exposed to radiation?" She stopped and grabbed up Caleb and hugged him to her, as if she could shield him from some terrible threat that way. "My babies—" Her voice was strained and hoarse again.

"I know. But, dammit—"

"But, *what*? We've got to get out of here!"

Dad turned on her with just a flicker of anger, quickly suppressed. "And go *where*? Have you thought about that?"

"The hills! East—out of the valley—it's only a few miles —maybe a hundred miles will be enough. But we've got to *go! Now!*"

Ephraim thought about the crowded highway. "Dad," he said quietly, "traffic's getting worse."

Dad turned to him as if surprised to hear something other than panic.

"I been watching out the window," said Ephraim. "Lots of cars. Trucks. Buses. Motor homes. Campers."

Dad nodded. "Hell." He dug his fingers into his scalp and seemed to be struggling to get things arranged inside his head. When he raised his eyes, he turned to look out the window at the trailer, then glanced at Mom.

She nodded. "It's all packed. The kids have been helping. Everything is ready, except to throw these last few things in."

As Dad, unwilling to go but unable to resist Mom's urgency any longer, bent to gather up the crib and load it, Ephraim stood silent for one more moment. Everything *was* ready. Besides packing all Mom's bags and boxes and suitcases, he had also gone quietly into his parents' closet and lifted down the holster that held his father's .38 revolver and a box of ammunition. He had slipped out earlier just for a second and now the gun and bullets were in Dad's pickup under the driver's seat, out of sight. He knew he wasn't supposed to touch the gun. But if the sheriff was busy and the police couldn't tell you what to do—or help you—who knew what might happen on a day like this? And the gun would be there, if it was needed.

Ephraim was used to taking care of things. Ahead of time.

35

CHAPTER 8

Ephraim dug his heels in and threw his weight into the pull; the big solid board gate leading into the yard from the driveway was heavy. As soon as the gate was open, Dad drove slowly through, inching the big heavy blue GMC ¾-ton pickup carefully between the gatepost and the garage. Inside the yard he wheeled over to the right, reversed, and started backing toward the trailer. Ephraim, following the pickup, could see Bones and Jocelyn standing beside the trailer. No animals were in sight, so he guessed that Turk and Kalijah were still safely hidden inside, along with Corkey.

Mom, he knew, was giving the house a last-minute check. She had always been nervous about fires, burglars, anything and everything, and wouldn't even go shopping without double-checking window locks, heaters, water faucets, and electric plugs. He could imagine how she felt now about leaving her house—her family's shelter and the only thing they owned of any real value—without even a neighbor to watch it. This only served to convince him of the extent of his mother's fear.

Dad was backing the pickup slowly toward the trailer, and

Ephraim hurried around to the pickup's left side so he could signal to him. When the ball joint on the truck's rear bumper was lined up directly under the socket on the end of the trailer's tongue, Ephraim made a sharp cutting motion with his right hand. Dad braked the truck and climbed out.

As his father cranked down the jack that supported the trailer tongue, Ephraim debated as to how much information he could get from his father.

"Dad?"

Grunt.

"Dad? Where—are we going?"

Dad screwed the trailer hitch down tight and then bolted and padlocked the safety chain. He was silent for several moments, and Ephraim was about to try again when his father stood up, looked at him, and said, "Ephraim—I got to tell you—I'm flying blind. There's nobody seems to be in charge of this mess. No plans, no orders. For right now, I can't think of anything better than what your mother says to do."

"But, Dad, the radio keeps saying they aren't going to have a meltdown. Not yet."

"Ephraim, things are out of control at Isla Conejo, or they wouldn't be doing all those emergency things they're doing. Firemen, police, doctors, talk about the National Guard, people coming out from the Nuclear Regulatory Commission. And when a nuclear power plant is out of control, there are liable to be—almost certain to be—radiation leaks."

"Why can't we—keep it out?" asked Ephraim. "What does the stuff look like?"

"It's invisible, Ephraim. You can't see it, taste it, smell it or feel it. You can get a fatal dose of it and not even know it. If there is a leak, it will be radioactive gases you can't see, or tiny drops of water in the air, or liquids that soak into the ground and poison plants and wells and rivers."

"But—can't we clean it up? Get rid of it?"

"That's the whole point, Ephraim. Radiation is the only thing you can't clean up or get rid of. It's like a poison you can't wash away, and it lasts forever."

"And . . . it's bad?"

"It kills people. Animals. Plants. Birds. Enough of it will kill everything."

Ephraim thought a moment. Then, "If it's that bad, how come they have it there? Didn't they know something like this could happen?"

Dad turned to look for Mom. "Well, everybody thought it would be safe and cheap, a good way to generate a lot of electricity. But lately your mother—and I guess other people too—have been reading up on this stuff, studying about it. Your Mom said months ago that Isla Conejo was dangerous and that people ought to do something about it—"

There was a *bang!* as the back door slammed and they looked around to see Mom staggering toward them. She was carrying Caleb along with her purse, a portable radio, a plastic jug of water, and a paper bag stuffed with something. As Dad ran to help her into the truck, he called back over his shoulder, "You kids ride in the back of the pickup. The shell will protect you, but be sure to keep the door locked. Whatever you do—*don't* open it till I stop the pickup and come back for you. Here—you can have the jug of water."

Bones and Jocelyn hastily scrambled into the back of the pickup, crouching quickly so they wouldn't bang their heads on the shell roof, and thumped down inside. The shell, or canopy, closed them in like a tiny house, and there were some boxes and sacks of supplies and a couple of old quilts and other gear strewn here and there. Ephraim took the water jug and as he prepared to follow them, he said, "Dad, how come those guys that are running the power plant let all this happen?"

His father had come back to slam the trailer door and make sure Ephraim locked it. Then for one moment he stood staring up at the sun, the pale blue summer sky, and the green world that lay spread like a flowered carpet as far as they could see. "I don't know, son. But when this is over, an awful lot of people are going to have to answer one hell of a lot of questions."

CHAPTER 9

The trailer creaked as it rolled down the driveway, and Ephraim could hear the purr of the pickup's engine. Dad was swinging wide to the left in order to make a right turn onto the country road that ran in front of the house. Ephraim crouched by the side window of the shell, his face pressed to the glass so he could see ahead at least a little bit. Bones and Jocelyn had not yet settled anywhere but were scuffling around, thumping their feet against the sacks of supplies and elbowing each other in the stomach.

"Sit down," grunted Ephraim. "Soon as we get out of the driveway, we'll be going faster."

Bones aimed a kick in his direction, and Ephraim started to grab for his ankle to pull him off balance. But just as he did, he caught a flash of red out of the corner of his eye. A red Ford pickup heading south on the county road had slowed to a halt directly in front of Dad's pickup and the driver of the Ford was leaning out of his window, yelling something at Dad.

Quickly Ephraim cranked the small window open so he could hear.

"—backed up all the way into town! I'm going to head south and then go east through Merced!"

Ephraim realized the other driver—Ray Fuller, whom Dad had tried to call—must be talking about the traffic on the state highway. It was always heavily traveled year round, because it was a straight line out of the Bay Area and into the Sierras. But today Ephraim had been watching and already knew that the congestion was far worse than he had ever seen it before.

Dad leaned out of his window and yelled back at the man in the red Ford. "Ray, have you seen Joe Foster, or Haskins, or Baker? I tried to call them to see if they needed any help, but I couldn't get an answer."

"Haskins and his wife are going with us—I'll pick them up as I go by. And Baker said he was leaving half an hour ago. You OK, Gabe?"

"God, I hope so! Got plenty of gas—food and water in the trailer. But what about your wife? Will she be all right?"

Ephraim saw someone in the other pickup lean forward and a pale face appeared over the driver's shoulder. Evelyn, Ray's wife, waved something blue. "Got my baby clothes with me!" she shouted. "In case I go into labor and he comes before we get back! Wish me luck!"

Dad stuck his arm out of his window and waved, slowly, in a long, descending arc; it reminded Ephraim of some kind of salute, a leave-taking between comrades being sent by different routes to a far-off war. Ray waved back.

Then the red Ford began to roll, picked up speed, and was soon just a red spot sliding between the walnut orchards as it headed south.

Dad, leaning out his window, watched it for one more

minute, and Ephraim heard him say, very softly, "God, You take care of them. And *all* of us."

Ephraim, staring down at his watch, could not believe that they had been waiting for fifteen minutes to get onto the state highway. There were three cars still ahead of them—two others had managed to squeeze into the bumper-to-bumper traffic, although one of them had narrowly missed getting sideswiped. A steady drone filled the air as the file of vehicles in the east-bound lane flowed past them like an unending river of steel and glass and rubber. It looked like every single person in the Central Valley was fleeing, some toward the west, but far more toward the Sierras to the east and south. There they would escape the prevailing winds from the northwest at this time of year, winds that would carry the radiation Dad and Mom feared if a major leak occurred at Isla Conejo.

When he looked out the small window in the side of the shell he could see cars, pickups, motor homes—some crowded with people but many having only two passengers, or even one. Ephraim had a brief, irritated moment when it occurred to him that if all the people driving nearly empty cars had taken on extra passengers like Ray and his wife were going to do, the traffic would have been cut by more than half.

He was going to say something about this when Bones spoke.

"I don't want to go to the hills today. It's not like when Dad gets his vacation. When it's vacation, everybody's happy —even Mom. And Dad talks about going fishing—" Bones broke off to stare at the small high window in the side of the pickup shell where they could see nothing but a blank stretch of empty sky.

Jocelyn glanced at her brothers. "Mom says we have to go. So *there.*"

Ephraim stared bleakly at Jocelyn. It would be nice, he thought, to be young enough still that you could believe your mom and dad were always right about everything and would always figure out what had to be done. *He* wasn't all that sure, anymore. . . .

The minutes passed like snails racing, and Ephraim watched as the three cars ahead of them, finally, one at a time, broke into the stream. At last they were first in line.

"Hang on," said Ephraim to Bones and Jocelyn as he got a good grip on the window crank to brace himself. "When Dad gets a chance to get onto the highway, he'll have to go fast. He's going to step on the gas hard—we'll get a jolt."

Bones stared at him indifferently. He and Jocelyn had gotten into a wrangle over whose elbow should be topmost as they sat side by side and scrunched together. They could easily have moved farther apart, but neither of them would give in and move first. Jocelyn was too small to win any dispute with Bones, but that had never stopped her from trying yet. Ephraim figured Jocelyn's life with Bones for an older brother would probably make her grow up to be a lady wrestler. She was already quick to punch and had come home from school several times with notes pinned to her shirt about her rowdy behavior.

For his part, Ephraim tried not to confront Bones any oftener than he had to, because he already knew that Bones could beat him. For that matter, Bones could beat almost anybody in a fight. Bones had punched out every kid in his own grade last year. It had finally occurred to Ephraim that if he—Ephraim—ever had his back to the wall, the best he could hope for would be for Bones to come into the fight on his side.

There was a sudden roar and a forward surge of the pickup as Dad seized his chance to break into the line of traffic on the highway. The pickup leaped forward. The trailer swayed almost off the right-hand wheel, and like three bullets the three children slammed down against the back end of the pickup bed. The pickup swerved left and right and then straightened out. They were on the highway.

Ephraim picked himself away from the door of the shell, glad that he had braced himself as well as he had. He had one moment of satisfaction that he had taken time to follow Dad's orders and lock the door—or they could have been thrown out and under the wheels of the trailer—and then he looked around at Bones and Jocelyn.

Neither of them had been braced as Ephraim had, and both had taken a bad jolt. Bones sat up. He looked dazed; his face, usually tough and hard as a baseball, was crumpled. He held up his left arm, and Ephraim saw a narrow line of blood streaking down from the elbow. He grabbed the arm to look at it. Bones was hurting too much to strike back, and Ephraim could see a wide, thick flap of the wrinkled elbow skin gouged loose and hanging. It must hurt something awful.

"Just a second," he grunted. "I'll see what I can do— Jocelyn? Are you— Hey, Jocelyn!"

Ephraim scrambled hastily past Bones, who was absorbed in his pain and the startling amount of blood that was beginning to pour out of his wound. Bones looked surprised—and scared.

Ephraim was scared too. But not only about Bones. Now he could see that Jocelyn's eyes were closed, and she was completely limp. He grabbed her around her shoulders and under the knees and pulled her away from the tailgate. She did not seem to be breathing. He started to put her down fast so he could go up and pound on the windows—call Dad and Mom—when he felt her catch a deep breath, and then her eyes opened.

For another moment she did not move or speak.

"Jocelyn—you okay?" cried Ephraim.

Her mouth worked as if she had forgotten how to speak. "The wa—"

Her chance sound made Ephraim think. He made a dive for the plastic jug of water Dad had given them. He unscrewed the cap, splashed some water in his hand, and let it drip over her face.

After a moment, as the water washed streaks across her dusty face, Jocelyn blinked again. Slowly, as if she was very weak or sick, she pulled herself up to sit, leaning on both hands. Then she saw Bones. "What happened to him?" she asked.

Ephraim poured more water into his palm and dripped it into Bones' wound. Bones winced, but for once he didn't seem to have much to say.

Ephraim screwed the cap back on the jug and started to take his shirt off. Luckily, he had worn a button-front shirt today instead of a T-shirt. As he started to rip some narrow strips off the tail of his shirt to wrap around Bones' arm, Ephraim said thoughtfully, "What happened? Oh, nothing much. This is just your normal wear and tear—happens every time—when you have a meltdown."

CHAPTER 10

"When are we going to get there?" Bones had asked the question twice before, and Ephraim was running out of answers, especially since he had no idea where they were going or when they would get there, either. Mom, he remembered, had said they had to go at least a hundred miles. But they were going so slowly—they couldn't have covered that much distance, or anything like it.

Bones and Jocelyn were huddled under the canopy's front window, facing backwards. Only by getting to their feet and crouching a little, could they have seen into the cab where Dad and Mom and Caleb were, although of course they could not have heard anything through the window. And neither could Dad and Mom hear them. So, as usual, both Bones and Jocelyn aimed their questions at Ephraim: "When are we going to get there?"

Ephraim had been looking out the side windows of the shell for quite a while (he was the tallest and could see best) and had for some time realized that they were not moving very fast. Houses, orchards, side roads were sliding past, but not at a great rate. They had been traveling now for nearly two hours, and Ephraim had only just begun to see the first low slopes that meant they were nearing the foothills. He remembered other times when they covered the same distance in half an hour.

Because of the trailer, he could see behind them only when the road curved and for some time now he had been aware of the fact that the line of cars following them stretched for miles. The line ahead was equally long. They were all bumper to bumper, almost as close together as railroad cars in a train. Every so often there was a squeal of brakes or a sharp blast on a horn as some driver miscalculated his hairline spacing.

Ephraim had never before seen traffic like this, and it scared him all over again. It had been almost impossible for Dad to get the pickup and trailer onto the highway. What if he had to slow down, turn off, or stop? With all the cars locked into this slowly moving steel snake, just one driver putting on his brakes to turn out could cause a ripple of collisions all the way down the line. Dad had said he had plenty of gas, but suppose something went wrong with the brakes? How good were the tires? If nothing else, the pickup's engine could overheat. What if—

"When are we gonna get there? Ephraim?" Bones reached out and tugged at his arm.

Ephraim stopped looking out the side window and crawled back over to Bones. Something in Bones' voice had struck him, and he wondered uneasily what it was.

Bones was crouched against the front of the pickup bed and the shell. In the shadowy light, his face was very pale and for some reason he looked smaller than usual.

Ephraim sat back against the shell wall beside Bones. "How's your arm?" he asked. He could see that the bandage was soaked with blood, but the blood that had run down Bones' arm was drying. Ephraim hoped that meant that the bleeding had stopped. If it didn't, he had no idea what to do, and the only other possibility—pounding on the window and signaling Dad to stop—seemed out of the question. Dad didn't have room to make any kind of a turn. And even if he did turn off the road—what then? Once out of the stream of cars, there would be the same fight to get back in.

"Bones," he said quietly, "I don't know when we're going to get there. But we got to look out for ourselves for now. Mom and Dad can't help us for a while. Does your arm hurt much?"

Bones, who had been sitting slumped over like a piece of paper left out in the rain, raised his head and looked at Ephraim. The younger boy's round face was pale and very sober. There were smears of dirt and blood on his sun-tanned cheeks and forehead; his mouth was trembling. And now Ephraim saw something he had not seen in a long, long time. Bones was crying.

Shocked, Ephraim reached out and awkwardly put his arm around his brother. Bones' body was big and hard, like a piece of timber or a sack of rocks. It felt funny to hug him —Ephraim had been defending himself from Bones for so long that he had nearly forgotten they were supposed to like each other.

And as they rode on through the heat of the evening, the smell of gasoline and oil, the rumble of tires on the roadbed and engines pulling, he realized that Bones was surprised too. Bones had always been so big and strong that he could hurt anybody he took a mind to, and no one could get back at him. Now he had found out something new.

He could get hurt too.

"When are we going to get there?" Jocelyn's voice was sniffly—she must have been crying. When Ephraim turned to look at her, he thought she looked like a stick figure, or maybe a scarecrow without enough clothes on. Her red bandana halter and blue plaid shorts barely covered her body, and her skinny arms and legs, all angles, seemed to have more sharp points than usual. The only thing about her that was big and round was the lump above her right eye. In its center was a small deep cut from which a little blood had trickled, and she had smeared the blood around, making her eye look bruised. Something about the eye seemed strange to Ephraim, but Bones was sitting between them, and he could not really see her well enough to tell what it was.

"Ephraim," said Jocelyn again, "when are we going to get there? I want—I want—"

"How about a drink of water?" asked Ephraim. You could usually get Jocelyn's mind off most anything by offering her something else—candy, a Coke, a ride on your handlebars.

But this time it didn't work. "When are we going to get there?" said Jocelyn again. She started to turn her head restlessly, as if it ached, from side to side, but then all at once she sat up. Before Ephraim could guess what was going to happen, she vomited.

Bones lurched sideways and Ephraim leaned over, grabbed Jocelyn's head, and held it as she retched. It was all he could do to keep from falling head first into the pool of slimy green vomit.

A second before Ephraim lost his balance, Jocelyn stopped gagging and sank back as if exhausted.

"Bones," said Ephraim quickly, "move over—so I can get her away from—yeah—like that."

As Bones crept toward the corner behind their father, Ephraim seized Jocelyn and dragged her away from the mess. "Hang on to her, Bones. I got to clean this up."

The smell of the vomit was overpowering in the small, hot, confined area. Ephraim had to clean it up, get it out, or they would all be sick. Then he caught sight of a grocery bag at the other end of the pickup bed. Maybe—?

He scrambled back to the bag—peered in. Yes—there was a roll of paper towels. He ripped off a handful, mopped up the mess as well as he could, and unlocked and raised the door of the shell just far enough to slide the wad of paper towels out. Then he slammed and locked the door again and crawled back to Bones and Jocelyn.

Brother and sister were clinging together like those hugging dolls you saw in the toy store, only these dolls were not smiling and happy.

Now Ephraim could see Jocelyn better, see her face and the right eye, the appearance of which had seemed odd a few minutes ago. Yes—there was a pronounced droop to the eyelid; and the colored part of the eye—he couldn't remember what it was called—seemed very large and black because the pupil was widely dilated. The other eye seemed normal.

Ephraim stared at Jocelyn for a long moment. A blow on the head. Vomiting. A huge red lump above the eye. And now the drooping eyelid, the oddly expanded dark part of the eye. All of it together made him think of a page in his health book at school last year—in the chapter on accidents and first aid.

And the page had been titled Concussion.

CHAPTER **11**

"Why is everybody else going the same way we are?"
Bones' voice was puzzled and a little uneasy.

Ephraim turned from the window in surprise. He had not
realized that Bones had been watching the traffic too. As a
rule, Bones seemed mostly concerned with things that were
within a few feet of him, but he had very strong feelings about
them. You could say, Ephraim thought, that Bones lived on
a very small hill, but there was no doubt that he was king of
that hill.

"There's too many cars," said Bones. "I never saw this
many before when we went to the hills. Why are all those
people going the same way we are?"

Ephraim wondered how much to say, especially since he
really didn't know a great deal about the situation himself.
But Bones was hurt—which was probably why he hadn't
punched Ephraim in the arm when he asked his question—
and Ephraim supposed that a person who had lost some
blood was entitled to know why.

"Well," Ephraim said slowly, "I've been listening to everybody. TV—radio—Mom and Dad. Close as I can tell, it's

like this. Isla Conejo—you know, that big nuclear power plant up north of Stockton"—Bones nodded—"well, they've had some kind of an accident there—"

"Like they blew up?"

"Not—yet. But . . . some of the machinery is busted, I guess. So it *might* blow up—"

"Well, if it hasn't blowed up yet, why is everybody going someplace? Why don't we wait till it does blow up?"

"Because then it would be too late."

"Too late for what?" asked Bones. "We've had earthquakes, and nobody went anywhere. And last year there was that big grass fire—remember the smoke? It hurt my eyes, but we didn't go anywhere. So why do we have to go off this way if the place didn't even blow up yet?"

"Bones"—Ephraim searched for words to explain something he couldn't understand either—"this thing is so bad— this radiation—that if we were still at home and it did blow up—well, we'd all die."

Bones thought for a moment. "Die. Like my rabbit died last year?"

"Yeah. Like that."

"But . . . we dug a hole and put Floppy in it and . . . put dirt over him."

"I . . . know." Ephraim hoped Bones wouldn't drill too hard on that aspect of the situation. He didn't think Bones was going to like the idea of somebody digging a lot of holes to put people in, and then covering them up with dirt. He was right.

"They can't do that!" cried Bones suddenly.

"But that's why—"

"No! Not holes!"

"Well, that's why we're all going to the hills—"

"And no dirt! No dirt on people!" Bones' voice was getting shrill. "Not on dogs and cats, either!"

Ephraim reached out and grabbed Bones' big fist, which was beginning to slash the air just under his nose. Calm him down, he told himself, quick, before he really gets rolling.

"Calm down, Bones," he said loudly, trying to make himself sound real convincing. "Keep your shirt on. There aren't going to be any holes. No dirt. See—that's why we're all going up into the hills. So we'll be safe!"

"Safe?" Bones stopped yelling. "Safe?"

"Sure. Safe. Why"—Ephraim decided quickly that if he'd ever have a good excuse for lying, he had it now—"we're all fine. Safe. Okay. No sweat. We're with Dad and Mom—even Kalijah and Corkey and Turk are okay."

"You're . . . *sure?*"

"Positive. Now, why don't you just lay back and go to sleep. Or sing a song. Think about something nice. . . ."

Bones was quiet for a moment, and Ephraim thought he might be getting a little drowsy from the heat and the swaying of the pickup, but then he began to speak again, softly, as if he didn't care much if anyone heard him. "When I grow up, I'm going to have a farm. Chickens. Pigs. Horses. Like Delbert Arnold's dad."

Ephraim looked at Bones. He could not remember ever hearing Bones say what he was going to do when he grew up. The present had always been plenty for Bones, maybe because he was usually in charge of it.

"Pigs?" said Ephraim. "You don't know anything about pigs."

"Delbert's dad has a farm, and they have pigs. I've seen them. I saw some baby pigs born once."

Ephraim wondered if Mom knew about the baby pigs. But

53

of course she didn't. Wow—what a racket *that* would have been. "Oh, yeah?" he said noncommittally.

"They're cute when they come out, like—like cookies Mom squirts out of that gun thing."

He meant spritz cookies. "Oh. . . . What does Delbert's dad do with the pigs?" he asked, not really caring.

"Well, he feeds them and they grow up big. And then people buy them and . . . kill them . . . and eat them. . . ."

Bones became strangely silent. Then suddenly he put his head down on his knees and began to cry. "Everything dies. Everything—dies."

Ephraim sat hunched and staring out the back window. He felt so sorry for Bones. Bones' arm was hurt, the world was blowing up (or melting down), and he had just realized that pigs had to die. All of it on one day.

"Hey, Bones," he said, reaching out to nudge him, but easy. "Hey, tell me about the horses—"

He figured nobody ate horses. . . .

CHAPTER 12

The line of cars had slowed almost to a crawl. They never quite stopped, but they were going so slowly that Ephraim felt he could have walked beside Dad's pickup and kept up with it. He had tried to figure out what was causing the slowdown, but he couldn't see up ahead that well. Looking out through the front window of the shell, he could also see through the windshield of the pickup, but that just gave him a view of the back of the car directly ahead. In back, all he could see was the front of their own trailer. He could see a little off to the side, so he rode most of the time with his face pressed against the open side window.

As they rounded a long curve that lay like a loop of ribbon over the flank of a hill, Ephraim began to hear noises—people yelling, a dog barking. He strained to see.

Now as the scene slowly appeared he began to piece it together. Something must have happened in one of the cars up ahead—maybe a breakdown, or, more likely, some driver had rear-ended the next car in line. At the speed they were moving, Ephraim doubted that any great damage could have resulted, but there must have been some kind of upheaval because evidently somebody's dog had gotten loose from one of the cars and was running back down the highway. He was big—an Irish setter—and he kept dashing toward the line of cars as if trying to cross the traffic to get to the other side of the road. He had a leash still clipped to his collar, and it was the leash that almost killed him.

Ephraim, seeing only a little but hearing shouts from the drivers ahead, guessed what had happened: the leash had caught under somebody's wheel and the dog, snapped short, had been struck by one of the cars. It hadn't killed him, but as the dog rolled away from the cars to the other side of the highway, Ephraim could see that he looked as if his left front leg was broken. The dog was howling.

Bones—who had been half asleep—snapped around at the sound. In another second he was at the window beside Ephraim. "What? A dog—I heard a dog crying!"

Ephraim was going to tell him what he thought had happened, but just then he saw some people rush out of a little house that fronted on the highway. The house had a fenced yard and a picket gate, and the gate slammed open as a boy a little older than Ephraim came running out.

The boy took in the scene in a flash and then sprang forward to catch the dog. The setter was howling and struggling—trying to get back to the car carrying his owner, which of course was moving slowly but certainly away from him.

As the boy caught the dog and held him—gently, Ephraim could tell that—the dog strained back in anguish. His long shrill wails cut through all the other noises like a knife. Ephraim thought there were probably some kids up there in one of the cars, far ahead now, who were looking back, and crying too.

Crouching, the boy holding the dog tried to wave, and Ephraim guessed he was signaling the kids in the car: Your dog's all right—I'll keep him till you can get back for him!

But just as Ephraim felt his sorrow for the dog and its owners begin to melt away, another person came out of the gate. It was an old man, gray and slightly bent. He was carrying a rifle in his hand.

And as Dad's pickup slowly glided up the shallow hill and around the bend, the scene was cut off, but not before they saw that the old man, after shaking his fist and glaring at the line of cars, had raised the rifle and aimed it at the dog.

Bones cried out as if he had been struck. "Is that guy going to shoot the dog? *Why?*"

Ephraim reached out and pulled Bones away from the window. "No—I don't know—" he said. "Don't look—"

Bones threw himself back down to huddle next to Jocelyn. *"Why?"*

Slowly Ephraim turned away from the window. "I—don't —know."

CHAPTER 13

"How come Caleb always gets to ride in front?" asked Jocelyn.

Ephraim looked around in surprise. He had thought Jocelyn was asleep—she hadn't talked or moved hardly at all for a long time. Bones had been quiet too, and Ephraim had crawled over to watch out the side window again, although it was getting harder to see much now because the sun had set and night was coming on.

They were into the real hills now, not just the long, slow-rising slopes beyond Oakdale, or even the little round oak-frosted mounds near Knight's Ferry. These were the high, rough, steep, brushy hills where the gold miners had dug and fought and gotten rich or starved to death back in 1849, where now the cattle herds grazed along dry streambeds mutilated by endless rows of tailings—heaps of gravel left from gold dredging. There were a few little towns here and there, and besides the big ranches there were lots of little places where somebody had a tiny lot with a few pine trees, a well, a garden, a mobile home. Clusters of these small holdings were strung along the road, and for some time

Ephraim had been watching as they passed the little settle-
ments, as their owners came out to stare, pop-eyed, at the
unbelievable caravan spilling past them. Looking at the
homeowners, Ephraim got an immediate impression that
they didn't look happy or welcoming. No one smiled or
waved the way they sometimes did in summer when long
lines of vacationers streamed up out of the hot valleys toward
the cool green meadows and forests. Today they looked
angry.

"How come Caleb always gets to ride in front?" asked
Jocelyn again.

Ephraim left the window—he was tired of looking at those
hostile faces anyway—and crawled over to Jocelyn.

But when he was closer to Jocelyn and Bones, he didn't
like the way *they* looked either. Bones was very quiet—in
itself something to worry about—and besides, his face was
flushed and his eyes looked funny. Ephraim reached out and
touched his brother's cheek, almost as if he thought the red
color was something that could be brushed off. Then he drew
back his hand in surprise. Bones' face was very hot. Fever,
Ephraim told himself. Mom would give him kids' aspirin,
but I don't suppose there's any back here. We didn't plan on
getting sick, or hurt, this afternoon.

He turned to Jocelyn, and she looked even worse than
Bones, although she did not seem to have a fever. Her face
was very pale, except for the red lump on her forehead and
the skin around her right eye, which was getting purple, as
if someone had socked her. Ephraim reached out to feel her
face but decided not to. As bad as it looked, it must be
awfully sore.

"Why does Caleb . . . why . . . does . . . Caleb . . ." Jocelyn,
staring straight ahead of her, did not seem to be talking to

anybody in particular. Or listening either, for that matter.

"Caleb rides in the front seat because he's the baby. You used to be the baby and you rode in the front seat," said Ephraim. "Don't you remember?"

Jocelyn stared at the floor of the pickup bed. "Why . . . does . . . Caleb . . . ?"

Ephraim sat very still. He was not a noisy person, not a screamer like Mom or a fighter like Bones, and, anyway, he had long ago figured out that yelling and punching never worked for him. Bones was allowed to fight, and Jocelyn to whine, and Caleb to cry. Ephraim was the oldest and Mom expected him to baby-sit Bones and Jocelyn and keep them safe in the backyard. When he grew up—if he ever got out of the backyard—he planned to go into the space program —NASA. He figured that you hardly ever took your younger brother and sister along if you were piloting a space station or driving a moon car around on Jupiter's third satellite.

But just now he wasn't in the cockpit of that spaceship. He was in the back of Dad's pickup in the middle of nowhere with ten thousand other people and two wounded kids. He decided it was time to try to let Dad and Mom know what was going on back here.

He got to his knees and pressed his face to the shell window that was just a few inches from the back window of the pickup. He didn't yell, but he beat on the window frame, a hard, regular, insistent drumbeat. Hew knew they could not hear his voice but could sense the vibrations of his pounding through the metal frames of shell and truck.

He thought they would never heard him. His knuckles were raw—Bones was crying again, Jocelyn gagging—by the time he saw Mom's head turn and he caught her eye.

He signaled frantically for a stop—pointed to Bones and Jocelyn—made signs they were hurt.

His mother stared at him. Only for a moment. Then her arms closed around Caleb, as if he was the only child she had left to hold on to. Her face was getting that look again, the one Ephraim called her hiding look. Then she shook her head as she huddled Caleb close.

Ephraim tried again—pointed desperately at Bones and Jocelyn and mimed their injuries. He was certain that she understood.

But Mom turned away, still holding Caleb, who was restless and wriggling, close to her. Ephraim could not tell if she meant they could not stop, or if she meant that what had happened to Bones and Jocelyn was not important.

When all he could see was the back of her head, Ephraim slowly sank back to the floor. And as he sat down, he contemplated the fact that Caleb wasn't bleeding. He wasn't scared. He wasn't even crying.

"Yeah . . ." he muttered. "How *come* Caleb always gets to ride in front?"

CHAPTER **14**

A little later, maybe fifteen minutes or so, Ephraim began
to think he was hearing something—voices?—from the cab
of the pickup. He rode for several moments without saying
anything about it, or getting up to look through the window.
At first he didn't know why he didn't want to look, but after
a moment he realized that it was because he guessed that
Dad and Mom were fighting. Otherwise their voices could
not have been loud enough for him to hear.

He glanced over at Bones and Jocelyn. As nearly as he
could tell in the poor light, they seemed to have gone to sleep,
huddled together like two puppies. It was quite dark now and
he couldn't see much when he looked out the side window
of the shell. The hills were great black bulges against the deep
gray sky, and the road, when he could catch a glimpse,
rounding a curve, was a chain of slowly moving yellow head-
lights behind them and red taillights ahead of them against
the steep dark slopes and the pine forest.

Suddenly the noises from the front seat got louder. There
was a shrill sound—Caleb was crying too. Ephraim winced.
He didn't really like to hear anyone cry—even somebody
who always got to ride in the front seat.

At last Ephraim turned over and stretched up to look through the window. Wondering was worse than knowing.

But at the same moment there was a sudden powerful swerve to the left. Ephraim barely had time to grab for the rolled steel edge of the pickup bed, as Bones and Jocelyn, limp with sleep, crashed into his chest and stomach. Ephraim's arms went around them, and all three rolled to the right side of the pickup. But there was an old quilt there that broke the force of their fall and they tumbled apart, roughly awakened, jolted, frightened, and yelling all over again.

Ephraim was the first to realize they had stopped. And as he untangled himself from Bones and Jocelyn, he knew too that Dad had not pulled off the road to the right, the way a driver usually would, but to the left, across the oncoming traffic lane. This would make it very hard for him to get the truck back into the eastbound lane again.

But the pickup had barely stopped rolling when Ephraim made a dive for the shell door and threw himself down over the handle.

"Let—me—out!" screamed Bones, throwing himself toward the door of the shell.

"Mama! Mama!" cried Jocelyn. Her voice was weak and she could barely move, but she was creeping toward the door.

"Wait!" shouted Ephraim. "Wait! Dad said never, *never* get out till he comes back for us—he might not really be stopped—we could get run over by the trailer—"

Bones, hurt and frightened more than he had ever been in his life, lunged at Ephraim. And he was so big and strong, even hurt and half-sick as he was, that Ephraim felt his hands torn from the lock.

"Wait! Wait, Bones—" he cried.

Bones was sobbing. "Mama—my arm—let me out—"

Desperately Ephraim reached behind where Bones couldn't see and struck his brother—not hard—on the injured elbow. Bones screamed and drew back.

And in that one second that he gained, Dad jammed his foot down on the gas and the pickup and trailer surged forward, jolted over rough ground, braked to a halt.

"See?" whispered Ephraim unnecessarily.

Bones and Jocelyn stared at him. Then both of them sank to the floor, waiting.

A second later they heard the pickup door open, and then Dad was standing at the shell door, looking in.

Ephraim turned the lock and pushed the door open. "Dad," he gasped. "Dad—"

"Daddy!" screamed Jocelyn as she tried to crawl to him. Bones made no sound, but he crept, shaking, toward his father with a look on his face that made Ephraim want to cry.

Their father, standing on the ground by the open door, spread his arms and closed Bones and Jocelyn to him. They pressed against him as if just the feel of his shirt, his whiskers, the smell of him could make them feel better.

Then his father reached out one hand to Ephraim, laid it on his shoulder for an instant, brushed his fingers briefly down Ephraim's arm. "Good job," said Dad. "You kept the door locked till I stopped and came back for you. I trusted you to handle it."

Ephraim reached out, too, and his hand brushed his father's arm. "Yeah. You too. I trusted you too," he said.

They were stopped at some kind of a wide place beside the road. About twenty feet away was a very small old-fashioned filling station, with the lights out and a Closed sign on the door. Behind the station were some cabins, probably an old motel. They were dark too, but there were a couple of dogs running around and barking at Dad's pickup, and Ephraim thought he saw a figure in the dim light—a man?—watching them too.

After Dad had let them get out of the back of the pickup, Bones and Jocelyn had wobbled up to the cab and crawled in with their mother and Caleb. Ephraim could hear them telling her all about how they got hurt, and how mean Ephraim had been to them. He could not hear Mom's answers, but he supposed she was sympathizing with them.

Ephraim thought as Bones and Jocelyn ran to Mom that maybe he ought to go and talk to his mother too, and tell her how scared he had been, and how he tried to take care of Bones and Jocelyn. But as quickly as the thought came, it faded, as he remembered that the little kids always got all the sympathy and he—the oldest—always got all the orders.

So instead of going to talk to Mom, he found himself standing beside his father, who was staring at the filling station and the dark cabins. The little kids, he told himself, can talk to Mom. I'm going to stay here with Dad while I've got the chance. Find out what he's going to do.

He wondered why Dad had stopped. He was sure it was not because Bones and Jocelyn were hurt. He was certain Mom had understood him, but he didn't think from the way she had acted that she had told Dad. She was the one who had said they had to run—head for the hills. She wasn't likely to tell Dad to stop.

"Dad, did you stop because we've gone far enough to be safe? Mom said we had to go at least a hundred miles—"

"No. We've only gone about eighty miles."

"Are we out of gas?"

His father seemed to be taking a lot of time sizing up the

filling station. Maybe he had seen something Ephraim had not noticed. Ephraim was about to speak again when his father answered, quietly, as if his mind was on something else.

"No. We've got plenty of gas. I still have a third of a tank left, and the auxiliary tank is full."

"Then why'd we stop at a filling station? Even if it *is* closed?"

His father drew in a deep breath and let it out, whistling, through his nose. It was a sure sign he was nervous, worried. "Ephraim—" he said, and then stopped, as if he was thinking over what he had to say. Then he started again. "Ephraim —I want to get out of that mob. When a bunch of people get scared and start running, they get dangerous. Running—this way, at least—isn't smart. It's crazy. And a person is just as dead whether he gets killed in traffic or by radiation."

Ephraim looked at him thoughtfully. Instead of telling Mom what had happened to Bones and Jocelyn, maybe he should tell Dad. He had a vague feeling that this was going to put him in a different position than the one he was used to, but for now he couldn't see anything else to do.

"Dad," he said, "both Jocelyn and Bones are hurt. I told them to brace themselves, but they didn't do it. And when you pulled out onto the highway, the pickup jumped forward so fast that they were thrown against the back end. Bones' left arm is cut bad—a big piece of skin is ripped back on his elbow. He acts sick. I think he's got a fever. And Jocelyn hit her head. There's a big red bump over her right eye. She's been throwing up. She acts funny. And they're both scared half to death."

He had thought that his father might get mad about Bones and Jocelyn being hurt, likely blame him for letting it happen. But he was not prepared for the trembling rage that overtook his father now.

Dad's hands clenched. He whirled suddenly, strode over to the pickup, and wrenched the door open. "Carrie!" he cried. "Look at these kids! Christ Almighty—what are we doing to these kids?"

Since the dome light was out, the pickup cab was very dark except for the constant on and off flash of car lights passing behind them. Even so, Ephraim could see Mom sitting against the door on the passenger side, both arms wrapped around Caleb, who was squirming and crying. Bones sat next to her, and Jocelyn beside him, near the steering wheel. They seemed to be watching Mom, expecting something—a hug? a kiss? comfort?—but Mom just stared straight ahead out the windshield.

"We have to run. We have to go—head for the hills," she said tonelessly.

Dad stared at her. "Carrie—look at Jocelyn. Look at Bones. My God—they're hurt! They need help—"

"—run. Head for the hills."

Ephraim thought she sounded like a stuck record.

His father leaned into the cab. His voice was urgent. He was trying very hard to be calm, but Ephraim could see his control was shaky. "Carrie, stop saying that. Listen—we've got to stop and *think* here—"

"—hills—"

"We've already got two kids hurt by heading for the hills. We've got to stop and think what to do—what's the *best* thing to do. Running—may not be—"

"Got to *go*. Get away—" Mom was rocking back and forth on the seat. She would not let go of Caleb, and she would not look at Dad.

His father stared at her for a long moment, and Ephraim thought he had never seen him look like this before. It was like Dad had never known Mom before, that she was a stranger, an alien.

Slowly Dad stepped back from the pickup, but as he turned away, he brushed his hand over first Bones' and then Jocelyn's heads. "Hang on," he told them quietly.

Then he turned to Ephraim. "Listen," he said, "we're going to go back there into the trailer and fix something to eat. Bandage up Bones' and Jocelyn's hurts. Make some coffee. And *think.*"

CHAPTER 15

The trailer was like a tiny island, lighted and smelling good as coffee perked and a big pan of canned soup heated on the butane stove. On each side of the table, Bones and Jocelyn sat with their backs against the front wall. Caleb was under the table playing with some paper cups. Ephraim had put Corkey's cage on the couch and he was flinging birdseed hulls through the bars as he ate. From time to time he rattled his water cup until Bones thought to fill it.

Mom had let out a yell when she saw Corkey in his cage sliding around on the trailer floor, and an even louder yell when Bones opened the closet door and Turk and Kalijah leaped out onto the floor. But she was so mad at Dad for stopping that she just flung herself down on the bench beside Bones, and sat there with her hands clenched to her mouth, her eyes shut.

Ephraim ran outside and scooped up a pan of dirt for Kalijah and shut him into the bathroom with it. They heard busy scratching noises for several minutes and then a questioning "Meow?" as he asked to be let out. Dad put Turk's leash on and walked him around the trailer and then brought him back in.

Coming back with Turk, Dad had left the trailer door open for air but closed the screen door, and every so often one of the dogs from the cabins came to rear up on the folding steps and peer in and sniff at them. Some seemed friendly, but one small dog was edgy, and they could see his sharp white teeth when he growled. Ephraim had clipped Turk's leash to the handle of the closet door. Turk now sat, trembling a little, alternately sniffing the soup and growling at the dogs. Jocelyn had Kalijah tucked into the seat corner close to her, and their two faces looked out at the rest of them. Ephraim thought they looked a lot the same—little and frightened and easy to hurt.

Dad ignored the dogs outside as he stirred the soup. He had taken charge of supper after he asked—no, ordered—Mom to calm down and take care of Bones and Jocelyn.

Gritting her teeth and with hands that shook, Mom got up and located her first-aid equipment, which was scattered through three or four sacks and cupboards, and then cleaned and bandaged Bones' arm. She was trembling so badly, Dad had to close the tape and bandage boxes for her. She did not take Bones' temperature but must have realized, as Ephraim had, that he had a fever, because she gave him some children's aspirin.

But when she turned to Jocelyn, Mom's face took on a hopeless look. All of them could see that there wasn't much that could be done for her out here, although Mom managed to crush some of the ice cubes she had earlier packed into a Styrofoam cooler and wrapped the ice in a towel, for Jocelyn to hold over the swelling on her forehead. Jocelyn was listless and hazy. Mom kept telling her to lie down, but she insisted on sitting up, with her arms wrapped around Kalijah. The only thing Jocelyn had said since they got into the trailer was, "How come . . . Caleb . . . always . . . rides . . . in the front seat?"

When she had taken care of Bones and Jocelyn, Mom sat down on the couch and huddled over, with her head down and her hands clenched at each side of her face. From time to time she whispered, "We mustn't stop. Got to *go*. Got to go *farther*—"

Dad did not answer her, but once he put his hand out and massaged the back of her neck, as if he knew that it hurt, and he was sorry.

Then he opened the cupboard over the stove. "I'm going to put the soup into mugs. Ephraim, get everybody a spoon. Here, Carrie, drink some coffee."

She took the coffee and looked at it and at Dad, but said nothing. Then Caleb, who had been playing on the floor, pushed his paper cups away and crawled out from under the table. Mom set her coffee down and got up and fixed a bottle for Caleb, and then sat down again with Caleb on her lap. Caleb's plump little hands curled around the bottle as he settled against her shoulder.

Dad reached up and turned on the little portable radio they always carried in the trailer, and Ephraim recognized the voices of Leonard Woolly and Barbara O'Neal, who were the regular newscasters at station KRAK in Sacramento.

"—have stated, Barbara, that the levels of radiation present in the steam that has escaped so far have not been critical. However—"

"Then they *do* admit that radioactive steam is escaping!" said Mom suddenly.

The voice on the radio continued. "Some officials here said that it has, and others deny it. There is *no* accurate, detailed information being given out. And there has still been no official statement from the governor's office indicating the need for evacuation."

The newswoman's voice was urgent. "But the California Highway Patrol has already reported that thousands— maybe tens of thousands—of people living in the Central Valley who were living downwind from Isla Conejo have already left their homes. They are fleeing into the Sierras to escape any radiation which could possibly be carried toward them by the prevailing winds. The California Highway Patrol says the roads into the mountains are snarled with traffic. They have called up all reserve officers to handle the situation. Some looting of abandoned homes has already been reported, and it has been suggested once again that the National Guard be called out."

"Yes, Barbara, people are not waiting for an evacuation order. But those who are fleeing will certainly be faced with problems of overcrowded campgrounds and, very likely, shortages of food and drinking water—"

"So why didn't somebody think about all this a long time ago!" cried Dad suddenly. "Why did they all wait till something like this happened?"

Mom raised her worn face to look at him. "Why did *you* wait?" she asked wearily.

"—stay tuned to this station, ladies and gentlemen, for further information. Our news-gathering crew here at KRAK will remain on duty around the clock until the emergency is over, and will release immediately any and all information we receive—"

"And no two reports will ever say the same thing," said Dad wearily. "Well—let's eat something. Maybe we'll all feel better." He took an extra cup and dipped soup into the mugs. Ephraim handed them around, each with a spoon. Bones,

72

who usually had a good appetite, picked up his spoon and started to eat. Jocelyn just stared at her mug.

As Ephraim took his first spoonful of soup—it tasted good and he was surprised to find that he was quite hungry—he heard some slight noises outside and supposed the dogs from the cabins were snuffling around.

Dad did not seem to notice the noises. He was watching Mom. "Eat your soup," he said tensely. "We need to—"

There was a sudden rattle of sound, and the screen door of the trailer was jerked open. A big man loomed in the darkness, and before any of them could move or speak, he climbed into the trailer.

Mom screamed—a faint little choked off sound—and Dad scrambled to his feet. "Who the hell are you?" he demanded. "What do you want?"

The intruder was a very big man, older than Dad, maybe forty-five or so, and dressed in denim overalls and a red plaid shirt. He had gray hair and a jet black beard, and his face was twitching as if he were very angry.

"This's my place!" shouted the big man. "What're you doin' here? What are all you crazy people doin'—comin' up here like this? Christ—thousands—thousands of you! Road's been jammed all day—people runnin' wild! I want you to get out of here! Now!"

Dad stared at him. "We aren't hurting anything," he said. "We just parked here to rest a minute—"

"Get out!" yelled the big man. "Get out! Get goin'!"

Mom was whimpering, with both hands pressed to her face. Bones and Jocelyn were completely silent.

There was very little room in the trailer, but Dad managed with only the slightest motions of hand and body to put Mom and all the children behind him. But that put him very close

73

to the big man. "Listen, friend," said Dad quietly, "we'll be out of here in just a few minutes. Just let us—"

"We don't want you up here! You're all runnin' wild—a stampede! Go back where you came from!"

"Haven't you heard? There's trouble at Isla Conejo— that's why we're here—why all these people are here—" Dad gestured at the endless stream of cars. "We had to run—get away from the valley—"

The big man's hands began to curl and Ephraim had a feeling he was getting ready to attack. The stranger's eyes were wide and staring, and his mouth hung open, like he was going to cry or scream. Ephraim realized suddenly that the man was not so much angry—he was scared.

Dad was carefully balanced, his legs slightly bent at the knees. Very slowly his arms bent too, and his big hands formed into fists. Overall, he didn't move very much, but it was enough.

The big man shifted, began to back toward the door. "I want you out of here," he repeated, but he was backing up.

Dad followed him, and as the stranger stepped backward out of the trailer, Dad was only inches from him. Ephraim, after one agonizing moment, slipped through the door after his father. He wanted to stay inside with his mother and the little kids—but something compelled him to go out and stand beside his father.

Outside, on the hard-packed gravelly stretch of ground, the stranger stopped, and Dad stopped too, beside the pickup. Dad opened the door as if to get in, and then turned. But as he opened his mouth to speak to the threatening man, there was a rush of sliding tires, and an old station wagon heading west out of the mountains squealed to a halt beside them. Four men piled out of the station wagon and clustered behind the big man. All of them seemed to know each other, and as Ephraim slid into a shadowy spot, just behind his father and the open pickup door and close to the pickup's seat, he suddenly was a lot more scared.

Dad was watching all of them carefully.

"What's goin' on, Dave?" asked the driver of the station wagon. He was a squat, bushy-haired man in some kind of striped coveralls. He smelled strong and was unsteady on his feet. "Is this some of that mob comin' up out of the valley?"

Dave—the big man—stiffened and his head tilted back. He glared at Dad. "Yeah. This guy—he's stopped here—trespassing—my property—"

"I told you—" Dad began angrily, but the man called Dave cut him short.

"We don't want *none* of you stoppin' here. You got to go someplace? Then *go*. Don't stop here!"

"I'm just trying to—" But once again his father's voice was cut off and covered by a babble of men's voices. None of them made sense, and Ephraim had a strong feeling that most of them were drunk. One of the newcomers drifted closer to stand directly behind Dave. In his right hand he carried something that gleamed faintly—a small axe, Ephraim now saw. He kept hefting and swinging it as he glanced restlessly around, as if he was looking for something to try the blade out on. Behind them all, Ephraim thought he saw something long and slender, like a rifle barrel, sticking up as if they had a gun standing on the front seat of the station wagon.

The man with the axe moved restlessly. "Bunch of crazy people," he grunted, with a nod that took in the family, the truck and trailer, and the endless stream of traffic. "Comin' up here where they don't belong, trampling decent people underfoot—"

"Yeah! Yeah! That's right, Frank!" The three other men,

faceless in the poor light, made Ephraim think fleetingly of TV movies of the Old West when the lynch mob gathers in front of the jail. You could turn off the TV, Ephraim told himself, but he didn't think there was going to be any way to turn these guys off. And they were shuffling closer, closer behind Dave, as if they were pushing him on . . . to something. . . .

"Tryin' to barge in here, take over—" the bearded man was advancing again.

"That's right, Dave!"

"Who do you think you are that you can come up here—run all over our land? Some kind of big bossman, a rich guy—"

"Yeah!"

"Rich—" Dave's eyes flickered. In the rhythmic on-again, off-again light, they could see him jerk his head slightly, taking in Dad's blue pickup and the Wilderness trailer, its door still standing open so the butane light fell like a golden rug on the ground beneath the folding steps. They could even smell the coffee and the soup. "Rich—" said Dave again. "Yeah—*rich*—"

"I'm not—" began Dad, but Dave cut him off immediately.

"Rich man—got more than his share. Now, you take a poor man like me, I ain't got no nice new pickup—"

"Watch it. Watch it, man." Dad's voice was getting very soft, very edgy. Ephraim felt sweat standing on his back.

Suddenly Dave thrust his head back. "Say, rich guy, how about we make a deal? You want to come up here—okay, you're here. I need a pickup—so I'll just take—"

"Sheriff!" said Dad and his voice was like the crunch of a steel pipe on somebody's knuckles. "You lay one hand on my truck and I'll have the sheriff on you so fast—"

They all seemed to think this was funny. Dave roared and slapped his sides. "Sheriff?" bellowed Dave. "How you gonna call him—on *my* phone? An' even if you could call him—he's on the *other* side of that road! You couldn't get the sheriff here for hours—days—"

Dad was silent. Ephraim was so close he could feel his father's body tremble. He himself was shaking so bad he could hardly stand. Help—help—he cried inside himself— we've got to get help—

"Why, I could blast you to kingdom come—" cried Dave, and he made a silly make-believe gun with his right hand, forefinger extended. "I got friends here to back me up! So, bang! You're dead! Now, I'll just take over that nice pickup—" He started to advance, and it was like the devil himself on the move.

In that instant, Ephraim remembered the gun he had hidden under the pickup seat. Bending closer, he reached very quietly in under the seat. Moving very carefully he slid the .38 out of its holster and just as carefully reached out and pressed it against Dad's right hand. Since they were both standing between the open truck door and the cab, they were partly hidden from Dave. Ephraim heard his father suck in a little quick breath, and then his big hand closed over the gun. He raised the gun till it was pointed at Dave's chest.

There was just enough light for all of them to see the gun.

Dave and the four men froze. None of them breathed, moved, spoke. Even the man who had been swinging the axe was motionless. The man standing nearest the station wagon turned ever so slightly to look at the rifle.

"No," said Dad very quietly. Then, just as quietly, he said, "Ephraim, go tell Mom and the kids to sit down. Lock the trailer door and come back here. Fast."

Ephraim had never moved so fast. In less than a breath he

was back. Dad pressed something—the ignition key—into his hand. "Get in. Start the motor."

Dad stood rock still, the gun like a black arrow pointed at Dave's heart. Ephraim scrambled into the pickup, felt for the ignition, put the key in and turned it. He had to slide forward to get his foot down on the gas but in a moment the truck's motor roared into life. He grabbed the light switch and pulled. Strong beams of light sliced the dark night.

Slowly, carefully, without ever losing control of his aim, Dad climbed sideways onto the truck seat. He slammed the door, gun still ready.

"Shift," he said to Ephraim, and his eyes never left Dave's face.

Ephraim had never been allowed to start the pickup's motor or touch the gearshift before. I've got to do this right, he thought. No mistakes. Carefully he pressed the gearshift lever till the pointer touched "drive."

Dad jammed his foot on the gas, spun the wheel over, and, in a heart beat, with tires throwing gravel, the truck surged into a tight turn, with the trailer bouncing crazily behind it and was speeding down the highway.

Westward. Toward the valley.

I hope he knows what he's doing, Ephraim said to himself, gripping the seat as he stared out into the darkness. I sure hope he knows what he's doing.

CHAPTER 16

The lights of Dad's pickup made a steady glow ahead, even though the pavement hardly got dark in the few seconds between oncoming, eastbound, cars. Ephraim felt funny being up here in the front seat with Dad while Mom and the little kids were in back in the trailer, but he sat hunched and silent, and did not speak even when his father reached over and handed him the gun. Then he slid off the seat, crouched, and pushed the .38 back into the holster under the seat on Dad's side. So far, Dad had said nothing since the moment he had rammed his foot down on the gas, and they roared away from the dark buildings beside the highway and the man called Dave. Ephraim was grateful for the silence, but he knew it wouldn't last. It didn't.

"I thought I told you never to touch my gun. Your mother's the only one allowed to take it down off the shelf, and then only if somebody breaks into the house when I'm gone."

Ephraim had time to think of several things, but he picked only one to say. "Dad, I'm safer right now with the gun than Mom will ever be—" And then he immediately regretted it. Dad didn't allow talk like that about Mom.

He waited for the blow to fall. When it didn't, he cranked his head around to grab a quick look at his father.

Dad was driving as if every muscle he had was operating at maximum strength, but his face hadn't even a flicker of expression. After a moment, Ephraim realized that this was one of those electrifying moments when somebody said something that was absolutely true, but it couldn't be acknowledged for some powerful reason that would shake up the whole universe. On the other hand, it couldn't be denied. So it was just allowed to stand.

Ephraim turned back quickly and fastened his gaze on the road ahead. Cold sweat poured down his back. When he said it, about Mom, he knew it was true in his head, but now that Dad had not punished him, he also knew that it was true in his heart. And he was terrified. It was bad enough to be in charge of Bones and Jocelyn. He didn't really want to be in charge of a gun, too.

They drove at least three or four miles in total silence. Dad tended the wheel and Ephraim huddled in the middle of the seat, afraid to creep closer to his father for comfort, in case Dad pushed him away, and afraid to creep over and lean on the armrest, for fear Dad would think he was trying to avoid him.

At last, when Ephraim was almost at the breaking point, Dad spoke. "It was you who brought the gun, wasn't it?"

Ephraim nodded.

"I thought so," said his father. "Your mother, scared as she was, would never have thought of a gun. Why did you do it?"

Ephraim kept his eyes on the line of car lights. He had known Dad would ask these things, and he had been struggling to get his answers ready. "Mom was scared," he said at last. "And everyone—police, sheriff—nobody told us what to do. Didn't even answer their phones. If you can't *call* the police, then what—" he broke off, then started again. "I thought . . . if it got really bad . . . we could protect ourselves."

Dad was silent for a while, and then he said, "This is what I was afraid of. When the law breaks down and there's no trained or responsible person in charge, in control—then some people get out of hand. Not everybody acts that way, but enough do to be dangerous. They panic, and panic spreads. They reach for guns. Axes. Times like these, people can get dangerous. Don't *ever* forget that."

"I won't forget."

Dad wiped his hand over his face. "Unload the gun," he said. "You've watched me—you should be able to do it. Then put it back in the holster and lay it on the seat. It's not legal to carry a gun that's hidden from sight. And it's not legal to point a gun at somebody, like I just did. Don't ever forget *that* either."

Ephraim took a deep breath, unclenched his fingers. As he slid off the seat again and reached into the darkness for the .38, he realized that somehow he had moved beyond that safe, backyard life inside the high fence. And now he wasn't sure if he really wanted to do this. It was scary out here.

For several minutes they had been listening to the Sacramento radio stations. Maddeningly, they continued to give conflicting reports.

One broadcaster stated that the governor had called out the National Guard, and he was immediately contradicted by someone who said he had not. Another interviewed a man named Walsh from Washington, D.C., who kept saying over and over that "the situation at Isla Conejo has been grossly exaggerated." Once he said, "It's under control now," and the interviewer pounced on the last word and tried to get him to elaborate. "Then the reactor *was* out of control?" But Mr. Walsh, whoever he was, managed to get off the air after talking for fifteen minutes, without having said one single useful thing.

Listening to the interview, Dad snorted. "You'd have thought they'd have somebody on the air who could tell us something. If there *is* anybody who really knows what the situation is. Why can't they at least broadcast some instructions—what to do—where to go!"

"I guess the ones who know all about it are out at Isla Conejo," suggested Ephraim.

Dad glanced at Ephraim. "I don't think anybody knows what to do," he said. He drove in silence for a few moments, and then said quietly, as if he were thinking out loud, "Carrie warned me that this could happen. I've got to give her credit —she's been reading the paper and watching stuff on TV. She's said all along that people should be paying more attention to these nuclear power plants. She wanted us to go and join some committee—something about investigating the safety precautions at Isla Conejo—but I was always too busy. Or too tired. And they kept telling us that we'd have cheaper electricity. Well—this is what I get for being too busy and too tired and too dumb . . . my family battered up and mixed in with a mob of refugees—"

Listening, Ephraim tried to think of something to say that would help, even though he realized that Dad didn't really expect him to. So he sat in silence, watching the oncoming cars. Then, ahead, through the trees banking a curve in the highway, he caught a glimpse of lights. Yes—there—as they rounded the bend he saw a cluster of small buildings, a parking area with lights, and a lot of cars, pickups, campers, motor homes, lined up in rows under scattered pine trees in the parking lot, and a few people moving around, or clustered in groups. There was a wooden signboard across the front of a small filling station and grocery store combined, that said Sierra Meadows Resort. Fishing. Groc. Hobart Boone, Prop.

"Hey," said Ephraim. "I remember when we passed this place on the way up. It looks nice, Dad. There're lights on —people look okay—just standing around talking. Why don't we stop here—hey, Dad?"

But even as he spoke, Dad's foot was on the brake pedal and the forward rush of the pickup and trailer began to slow. Ephraim braced his feet as the heavy hand of gravity pushed him toward the windshield, while he quickly scooped up details. Somebody was calmly pumping gas into a Toyota pickup while somebody else cleaned its windshield. A row of tiny houses behind the filling station had porch lights on, cars parked in driveways, fenced yards where a few people stood quietly talking.

Dad spun the wheel for a right turn, gravel crunched, and the pickup rolled to a halt between a green and white Winnebago motor home and a black van. He opened the door on his side and was out almost before the pickup had stopped rolling.

"I got to check on your Mom and the other kids," he said to Ephraim. "Keep your eyes open and don't get lost." Then he was gone.

Ephraim sat still for a moment and looked around. He still shook a little whenever he thought of the dark moments at the last stop when the man called Dave and his gang had threatened them. Would this place be like that? He wished for just one moment that Dad hadn't made him unload the .38. On impulse, he grabbed the gun and shoved it back under the seat. It made him feel better to have it there as a hidden surprise in case trouble should break out, even though it was now unloaded, in spite of the fact that Dad had plainly told him it was wrong to carry a concealed gun. Dave hadn't managed to steal Dad's truck, but they would never know what might have happened if the gun had not been there.

He opened the door on his side and slid slowly to the ground. He stayed very close to the pickup as he inched around to the front end where he could see better. He could hear the drone of the traffic on the highway, still heavy and flooding toward the east, and see the constant flash of car lights as they passed.

Behind him, Ephraim heard Dad unlock the trailer door and a babble of voices—and Bones' and Jocelyn's crying— as Dad climbed into the trailer. Ephraim wished he could go and see if his mother and Bones and Jocelyn and Caleb and the animals were okay, but Dad had told him to keep his eyes open. Torn once again between wishing to be among the little ones, protected and cared for, and a much different need to find out what was going on, and maybe even to help direct events a little, Ephraim found himself pressed against the front bumper of Dad's pickup, eyes and ears open, straining to see and hear whatever he could.

There were a couple of rows of parked cars and pickups between him and the service station so there was not much light back here, but he could see a little. Across the parking lot, a man was walking a dog on a leash, and somebody— he sniffed sharply—was making coffee in one of the campers. It smelled good.

To the right of Dad's pickup was a pine tree, and beside it stood a black van. It had one of those funny round windows and orange and yellow flame patterns painted on its sides. The door on the driver's side stood open and now Ephraim saw that the driver of the van was a young man with a blond beard and long hair, and he was talking to someone inside the van whom Ephraim could not see.

"Hey, man, it's a parade. Look at the parade! Ten million cars in one parade! Maybe it's the Fourth of July, or Cinco de Mayo. Is this Cinco de Mayo, Ben?"

Somebody—Ephraim guessed it was Ben—laughed and said, "Sure, it's a parade, Ron. It's all the holidays mixed up together. But we're the only ones celebrating."

"The only ones," said Ron gravely, "that are *happy.*"

"Yeah. Happy." Something—a beer can?—came sailing out through the van window and clattered to rest on the graveled parking area, and Ephraim, squinting in the dark, realized it was rolling down to come to a halt with a sea of beer cans beside and under the van. The van must have been loaded with beer when it got here, he thought, but it would go home empty, when the time came that they could all go home. If that time ever did come . . .

He was just about to turn away to find Dad and report to him that they had just moved into a bad neighborhood when suddenly he noticed something else. The driver's left arm was hanging out the van window at a rather odd angle. Edging forward a few feet, Ephraim got a better look, and was surprised to see that the man's hand was fastened somehow to a length of chain. And the other end of the chain was looped and padlocked around the trunk of the pine tree.

Son of a gun. The happy man was handcuffed to the tree.

Ephraim turned abruptly to go find Dad and fill him in on

all these strange facts, and instantly plowed into his father, who had come up behind him.

"Hey—Dad—" said Ephraim.

"Ephraim—come here—" Dad reached out and grabbed his shoulder.

"Yeah, but—Dad"—as his father propelled him toward the trailer, Ephraim pointed back at the van—"Dad, he's handcuffed!"

Dad stopped, jerked a sharp look back. "Yeah? *Yeah?*" He was silent for a moment, but Ephraim thought he had sounded pleased.

He was right. "Good," said Dad. "Some people ought to be handcuffed. We don't want a repeat of that mess with Dave and company."

"Right. But I wonder who tied this guy up?"

"Don't know," said his father. "Somebody here must be using some common sense—the guy's probably drunk or something. Look, you can see all those beer cans under the van. Anyway, I want you to stay here with Mom. Keep an eye on things."

At the door of the trailer Ephraim forgot all about the man in handcuffs. Dad had lit the butane light, and in its pale glow Ephraim could see inside the trailer. Bones was huddled on the floor with his arms around Turk. His arm had a clean bandage and his shirt had been changed. But although Bones still looked little and hurt—Ephraim was again surprised at how differently he saw his brother today, tonight—it was clearly Jocelyn who now most needed help.

Mom had made a place for her to rest on the couch, with her head on a pillow and a light cover over her. Jocelyn lay as if asleep, but her face was very pale, and her whole body quite still.

86

"Ephraim," said Dad behind him, "Jocelyn is hurt bad. You stay here with your mother and the kids while I go and see if I can find any place to take her for emergency treatment. There's got to be a doctor or a hospital somewhere. I'm going into this little store—I won't be long. You stay."

"But, Dad—the traffic—" Ephraim started to say. But before he could finish, his father had bolted off between the parked cars and disappeared into the little crowd of people around the station.

Ephraim wondered if he ought to climb into the trailer, but decided not to. He could see what was going on out here, and could yell for Dad if something happened.

Inside the trailer it was very quiet. Mom was kneeling on the floor beside Jocelyn, and now Ephraim could see she had made a fresh ice pack, which she was holding over Jocelyn's injury. In her other arm she held Caleb, who was quiet for once, sucking on a bottle of milk. Mom had managed to get his diaper changed and his shoes and socks off, and he lay back now, contentedly sucking his bottle and watching his toes wriggle.

"Mom," said Ephraim softly, "are you—okay? And Bones and Jocelyn?" He knew it was stupid to ask. Anyone could see they weren't okay, but he felt he had to say *something*.

His mother turned to him, and Ephraim thought he had never seen such a desolate, hopeless face. He knew he would remember it forever.

"No," said Mom quietly. "They're not okay. Bones is hurt. Jocelyn—my God!—my Jocelyn—I may have killed her. And only because I tried to save—us—all."

CHAPTER 17

Ephraim sat on the top step, his back to the inside of the trailer, watching. Dad had disappeared into the store, and until he came out there would be no chance to question him, so Ephraim kept busy sorting out the scene around him.

It was fully night now, and dark, but he was able to see that the Sierra Meadows Resort was a collection of old buildings, although they were not run-down or trashy. The parking lot, though it now held probably fifty or more assorted vehicles, was not littered, except for the nest of beer cans under the black van, and the few outdoor lights all had bulbs that worked. There was even a row of velvety red geraniums under the store's front windows.

The cluster of tiny houses that lay behind the store and service station looked as if they had been built by amateurs with bargain basement lumber, but they were all painted and each one had a small fenced yard, some trees and flowers. There were two or three dogs who were patrolling, stiff-legged and with bristly tails, just inside their own fences. The smallest dog of all, a tiny fellow with patchy brown and white fur, sat with his bulbous nose thrust through his wire gate, as if he were operating a Distant Early Warning system and at the first whiff of trouble he would let everyone know they were being attacked.

The motley crowd around the store, however, didn't seem to be very unruly. Ephraim noted that several women were clustered off to one side where their lighter voices wove an undercurrent of sound that he could hear during the slight breaks in the passing flow of cars. Several men had gathered in a group closer to the road and were talking too. Some had cans of beer, although Ephraim could not see anyone who appeared to be drunk, and some were smoking. After the terrible moments earlier this evening when they had been met with fear and anger and lawlessness, this place seemed like a different world—or the same world but run by different people. Somebody, after all, had handcuffed the man in the black van. That must mean that some effort was being made here to control chaos, and if so, maybe the effect would spread out, like ripples on a lake. He decided that he would find out, when he had a chance, who or what had made the difference here at Sierra Meadows.

There was a sound inside the green and white Winnebago that was parked to the left of Dad's pickup, and Ephraim's eyes flickered over there. Then the door halfway down the coach on the right side popped open, and a man got out, turned, and raised his hand to help a woman climb down.

Ephraim stared at them curiously. The woman was older than Mom, more like Grandma Dearborn, with tightly curled silver hair and eyeglasses so thick the lenses looked like round ice cubes. She was wearing a beige pantsuit. She walked stiffly, as if cramped from riding a long time, and, curiously, she carried a goldfish bowl in one hand. As the door of the Winnebago swung shut behind her, she said to no one in particular, "Horatio goes with me."

The man glanced around the parking lot. "You and your

goldfish," he said, and Ephraim thought he sounded a little annoyed. "Don't stumble. You'd both be hard to pick up."

Sitting barely six feet from the couple, Ephraim could not help hearing them, but to leave them their privacy, he carefully looked the other way.

"Why don't you wait here for a few minutes, Helene, while I go into the store and see what I can find out?"

"Will you be long, Ward?" Helene's voice was so calm that Ephraim marveled. She was the only person he had heard all day who sounded as if the world had not just turned upside down.

"Only long enough to make sure this is the right place."

The man turned and walked away, and Ephraim stole a quick glance. The lights of the station shone on the newcomer as he passed between cars, revealing a smooth head, bald, as Dad would say, as a billiard ball, and a short stout body clothed in a white turtleneck sweater, a leather jacket and gray slacks.

Ephraim thought he looked to be pretty well off. He wondered if having any considerable amount of money would make a difference in a situation like this, and decided to watch them and find out.

The woman—Helene—turned away from the Winnebago as calmly as if she were out for an evening stroll. A few steps brought her almost eye to eye with Ephraim. He thought she might glance away, like strangers do, but she didn't.

"Well," said the woman, smiling a little, "tell me, young man—will we find a way to keep the world from blowing up?"

"Mom, there's a lady out here. I told her about Jocelyn, and that Dad's trying to find a doctor. The lady wants to know can she help." Ephraim stood on the top step in the doorway of the trailer. He was purposefully blocking the entrance because he knew Mom would be upset—to put it mildly—if a stranger, even somebody trying to help, came up behind her.

Mom was sitting on the trailer floor beside Jocelyn, helping to steady the ice pack with one hand, and using the other to push toys close to Caleb, grab his shirttail if he ventured too near the door. Bones sat on the floor too, his back to the couch, but Mom did not touch him, or even seem to see him. Ephraim had a fleeting, hazy feeling that he had never experienced before, and the best way he could frame it in his mind was to say to himself, Too many of us. There are too many of us. Dad can manage all of us but Mom can't—

But he said aloud, "The lady says she used to be a nurse."

His mother's eyes suddenly focused on him. "Nurse?" she said sharply. "Where is she?"

"Right here," said the lady with the goldfish bowl. She was standing on the ground behind Ephraim. "Here, young man. Take my fish, please."

Ephraim turned and held out his hand for the little glass bowl. He didn't want to hold her fishbowl—he was sure he would drop it—but if Mom wanted to talk to her, someone would have to hold it and help her up the steps. As she now began to hoist herself up into the trailer, Ephraim leaned over and thrust the little bowl under Bones' nose. "Hold the fish," he said abruptly. Maybe watching the little fish—he *was* pretty—would take away the cold, lost look from his face.

Bones roused himself to reach out, a little surprised, and cupped his tanned, rough-knuckled hands around the bowl. As he looked down at the bright little fish, his cheeks warmed and he did look better.

It wasn't easy getting the silver-haired lady up the steps

and into the trailer. Her legs seemed stiff—maybe she had arthritis like Grandma Dearborn—and she had to climb the steps by grabbing the door to lever herself up. Ephraim put his hand under her elbow and pulled till his eyes popped.

"You're—a nurse?" asked Mom tensely. She grabbed Caleb, who had caught sight of the fishbowl and was trying to get his hands on it.

Bones, stirred out of his numb silence, gathered himself together, blinking as if he had just waked up, and rose to his feet, carrying the fishbowl to the table at the front of the trailer. There he sat down on the bench, put the bowl on the table and lowered his head and shoulders so he could stare closely at the little fish. They were practically eyeball to eyeball. Kalijah, who had been hiding under the table, suddenly emerged from the shadows and poured himself up onto the table in a smoothly flowing blur of speckled gray fur. In another moment his pointed face and shell pink ears hung over the fishbowl next to Bones. Turk, who was still tied to the table by his leash, appeared to sense that there was something new and interesting in the glass bowl, and with great effort he scrambled up onto the bench beside Bones. He then sat, quivering a little, with his whiskery muzzle just brushing Bones' cheek, as both of them stared at the little fish.

Kalijah, his tail switching, began to hum a soft, high-pitched, hunting song as his cat's nose told him "Fish!" But Turk only looked, breathless, and his bottle-brush tail scrubbed the back of the bench. Turk was the only living being, Ephraim thought, watching the little black dog, who literally loved everything and everybody on sight. Ephraim knew Turk would have licked the little fish's nose if he could, like he tried to do to the baby gopher snake they had found in the yard one day. The snake had hustled away like a string of excited parentheses, but the fish in the bowl seemed more inclined to engage in a meaningful, though prudent, relationship. And Caleb, seeing the shimmering little fish in his bowl, dropped his paper cups and went to look at it. When they were all absorbed in each other, Ephraim turned back to Mom and the strange lady.

His mother was on her knees beside Jocelyn, her hand and arm supporting the child while the lady eased a folded blanket in under Jocelyn's head.

"Yes, let's keep her head a little higher," said the lady. Then her right hand went to Jocelyn's wrist, where she tapped one finger into a certain little hollow. She turned up her other wrist and stood silent, frowning a little, as she stared at the face of her watch. "Weak," she said after a moment or two, shaking her head slightly. "Weak."

Mom raised her face to the lady, and Ephraim felt tears filling his eyes at the expression he saw there. She looked desperate.

"Mrs. Murdock—" whispered Mom, and Ephraim understood that his mother and the stranger had exchanged names—"What—? How bad—? God—what can we do?"

Mrs. Murdock stood silent for another moment, staring down at Jocelyn's white, still face. Although her expression was as blank as a closed door, Ephraim had a feeling that, like the door, it concealed something going on inside. For one brief moment Mrs. Murdock turned to glance over her shoulder, out the door in the direction in which her husband had disappeared. Then she shook her head.

"Hospital," she said abruptly. "She's got to get to a hospital right away."

"But—there's no—we're a hundred miles from a hospital," said Mom hopelessly. "And look at that highway—the traffic—" She broke off.

There was a momentary silence again, and then behind

them they heard a soft hiss. Absently Ephraim, his mother and Mrs. Murdock turned. Kalijah had raised a paw to see if the pretty fish could be caught, and as they turned to look, Bones hissed warningly at the cat, seized the predatory paw, and held it in his hand. Kalijah's eyes flickered green lightning, but he could not move with one of his velvet weapons under lock-down.

Mrs. Murdock stared at Bones, at the fish, at Kalijah, and at Turk, whose tail had never stopped wagging. She glanced down at Caleb, who was again trying to put a large paper cup into a small one, and her closed-door face changed slightly, eased, like a glimmer of light showing as the door opened. Then she spoke.

"Have to—find a way," she said briskly. "Here, young man, what's your name? You're Ephraim? Yes—all right, you come with me. You must help me find Ward—my husband. Fortunately for you—all—I just happen to have a trick up my sleeve. Come on, Ephraim. I'm about to pull a rabbit out of a hat!"

And as Mrs. Murdock grabbed Ephraim's shoulder and began to clamber down the steps, Bones looked up. "Rabbit?" he said hopefully. "Rabbit?"

CHAPTER **18**

Stumbling along after Mrs. Murdock in the darkness between the parked cars, Ephraim started to ask her if she could see where she was going. Then she halted abruptly, reached back to grasp his shoulder again and stationed him in a position just in front of her. "You'll have to see the way and lead me," she told him. "It's all I can do to push my feet, let alone pick out the way in the dark. And hurry. We must hurry!"

Ephraim groaned. Here he was again, thrust up out of the safe rear ranks and put to work leading. He couldn't see the ground at all. Ouch! There was a rock—and now this old lady had just made him responsible for guiding her to wherever it was she wanted to go. "Why me?" he muttered. "Why am I always doing things I can't do?"

There was a sudden sharp *plunk!* and Mrs. Murdock let out a cry. "Ephraim—wake up! Didn't you see that thing sticking out?" she cried, grabbing her head.

Ephraim whirled around and found that he had led the old lady smack into a low-hanging tree limb. He was shorter than Mrs. Murdock, so the limb was above his head, and he had not seen it at all. "Sorry about that," he grunted as he steered her around the tree. For just a moment he was grateful Dad hadn't seen that—Dad would have thumped him on the head with his knuckles for letting an old lady get hurt.

But now they were past most of the parked cars and emerging into the circle of blacktop that underlay the gasoline pumps and the little store. Here the light from strings of small bulbs was good enough so they could see well, the footing was smooth, and best of all, quite a few people were gathered in groups here and there. Ephraim was sure Mrs. Murdock would now find her husband, or somebody, and get on with whatever it was she planned to do. He was going to make a fast getaway while she made arrangements for her rabbit and her hat. He wanted to get back to the trailer, where Dad had told him to stay, before he got in trouble with his father.

But just as he spotted Dad coming out of the store, Mrs. Murdock's hand bit into his shoulder again. "Over there," she said eagerly. "I hear him—my husband. Help me, I've got to talk to him!"

"Why me?" sputtered Ephraim to himself again as Mrs. Murdock, now on solid ground, made her way toward a cluster of men by the gas pumps. Then he saw the bald-headed man from the Winnebago and breathed a sigh of relief. Mrs. Murdock would release him now, so he could go over and explain to Dad why he had left Mom and the kids alone after being told to stay there.

He started making tentative break-away signals to Mrs. Murdock, but she was now steering him toward her husband. "Ward!" she called out. "Ward! Over here!"

The bald man turned, saw Mrs. Murdock, and with a measured, assured step crossed to meet her. He seemed to be a person who took charge of things. Ephraim wondered if Mrs. Murdock had the same kind of control. Over rabbits. And hats.

Still leaning heavily on Ephraim, Mrs. Murdock halted and faced her husband. Surprisingly, he spoke first.

"Helene," he said, "I've been talking to the store owner, Hobart Boone. He's a remarkable person. You ought to meet him. When this mob"—he gestured at the highway, the parking lot—"when this mob started to arrive, he and his wife simply got things organized. No panic, no trouble. He's formed committees and put them in charge of pumping gas, selling groceries, even the rest rooms, so everybody gets what they need, but nobody grabs too much. He's even got a drunk handcuffed to a tree. He's a retired army sergeant, and I'll say this for him—he knows how to handle people. And trouble."

Mrs. Murdock nodded irritably. "Too bad there aren't more like him. Maybe if we'd had people of his caliber in charge of Isla Conejo, we wouldn't all be in this mess. But —what about Floyd? Is he here yet?"

Ward Murdock nodded. "He'll be here in a few minutes. Mr. Boone heard him on his CB radio and he's getting the field cleared." He gestured to a spot behind the little houses where there was a flat area about the size of a baseball diamond. There were some people hustling around it, doing something. "All my plans have worked out perfectly." Mr. Murdock smiled confidently. "Floyd will pick you up and take you to Reno, and you will be safe there at your sister's house, out of range of any trouble at Isla Conejo."

Mrs. Murdock was leaning about equally on Ephraim and on her husband's arm. Ephraim's shoulder was going numb.

"Ward, I do appreciate your setting up this plan to take

care of me," said Mrs. Murdock, "but I have to tell you—I'm not going!"

"What are you talking about—I want you where you'll be safe—"

"There's a child here who's hurt—"

"But—"

"Ward—an injured child has to come first. You know I've always believed that. So—*she's* going to Reno instead of me. Boy—Ephraim—find your father and tell him— We're sending your sister to the hospital in Reno. We have a special way to do it, and she can be on her way in just a little while."

So *that* was Mrs. Murdock's plan! She had some way to get Jocelyn to a hospital!

For a moment, Ephraim stood still as he felt something cold and hard dissolve in the pit of his stomach. Only then did he realize how scared he had been about Jocelyn.

"Oh, boy!" he gasped. "Oh, boy! Hey—there's my Dad! I'll go tell him."

At that moment Mrs. Murdock unlocked her hand from his shoulder and Ephraim lit out running toward Dad, who had seen him now and was just getting ready to rip into him. But as he crossed the intervening yards, besides being glad that Jocelyn would be cared for, Ephraim knew he had to find some good way to say it—about the hospital.

Not to Dad. Dad would be overjoyed to hear that help— of whatever kind—was offered to Jocelyn.

But Mom—what would Mom say when they told her Jocelyn was to be sent away—with a stranger—to a strange city? Worst of all, Ephraim already knew that Dad would send him to tell her.

And he did. As Ephraim gasped out his news, Dad

grabbed him and his face broke into the first smile Ephraim had seen there in hours. Then Dad shoved him toward the trailer.

"I'll go talk to the Murdocks—find out what they're going to do," he said. "You go tell Mom! And stay there till I come!"

"*Why me?*" bawled Ephraim as he galloped back across the parking lot. "*Why me?*"

CHAPTER 19

They were gone.

Ephraim stood in the open door of the trailer and stared at the empty couch, the table, benches, the sink from which a sodden towel hung, dripping water on the floor. It was the towel which Mom had used to hold the ice for Jocelyn's forehead.

Ephraim climbed into the trailer. Maybe they were in the bathroom? But it was too small to hold all of them at once.

He opened the bathroom door a crack. From the whines and scrabbles he knew immediately that Turk and Kalijah were in there. Bones must have shut the cat and dog into the bathroom to keep them safe. He would think of something like that. Corkey's cage, he now realized, was on the floor under the table, with something—an old sweater—thrown over it. Bones must have done that too.

Mom's purse lay on the couch, dumped over, with some of its contents, including her billfold and key ring, beside it. Seeing the keys gave him an idea, and he banged the screen door back, jumped to the ground, and ran up to look into the cab of the pickup. It was empty.

Ephraim walked slowly back to the trailer and stood for a moment frowning down at the graveled surface of the parking lot.

Why on earth would Mom leave the trailer, especially with the children? She would have to carry Jocelyn, and that left Caleb for Bones to lead or carry—and Bones was not in very good shape himself. Why hadn't Mom waited here? He was sure they had not gone toward the store—he would have seen them. They weren't in the pickup. Beyond the parking lot on one side there was nothing but the steep slope of the mountain clad in heavy forest, and on the other side, the highway.

Where could they go?

Bones, he thought, as he jumped down the steps, slammed the trailer door and started running. Bones is strong and he takes care of things he loves. I've got to believe Bones will stay with Mom, keep her and Caleb and Jocelyn together and all right. And I've got to find Dad.

The crowd had thickened around the gas pumps and the store—there were several new cars and pickups and campers now—and in the press he could not see his father anywhere. Looking back over his shoulder for an instant, he noted that the happy man in the black van was no longer handcuffed to the pine tree. He and his companion were gone and the van seemed empty. Maybe they were over there in the cluster of people milling around the store. He did not see the Murdocks either now, which surprised him, as they were people who would stand out in any crowd.

For a minute or two he moved rapidly, scouting each knot of people for a glimpse of any of them—Dad, Mom, one of the kids, or even the Murdocks. In addition to the fact that more people had arrived, there seemed also to be an extra buzz of excitement among them, but that only annoyed him because it made it hard for him to keep track of where he had searched.

Suddenly he saw something new—a black and white sedan parked just clear of the highway.

There was a gold insignia on the door, and he could just make out some words—*California Highway Patrol*—above it.

CHP! With the scene up the road when Dave and his ragged band had threatened to grab Dad's pickup seared forever into his mind, Ephraim knew that he'd never see anything more welcome than that black and white car. He started to run toward it, but braked to a stop when he saw that it was empty. Spinning between strides, he started back for the store, and then everything opened up at once.

There were Dad—the Murdocks—and over there, talking to some people, the CHP officer.

"Dad!" he yelled. "Dad—come here—I can't find Mom! She's gone!"

Dad reached out for him and he was smiling—hopeful, eased, relieved. I guess he's glad about Jocelyn, Ephraim thought.

But then his father realized what Ephraim had said. Dad's arm shot out, his hand closed on Ephraim's shoulder and yanked him close. "Ephraim! What are you doing here? I told you to stay with your mother—"

Ephraim staggered in his father's grip. "Mom's gone—and she's taken the kids with her! I've been looking for them, and I can't find them anywhere!"

"Gone? Where?" Dad's face turned icy. "How could she leave the trailer—she couldn't carry both Jocelyn and Caleb!

Oh, God"—he broke off and punched his fist against the gas pump—"what the hell is that crazy woman up to now? All she had to do was wait there for me—only a few more minutes, and we could have had Jocelyn on the way to a hospital—"

"What is it? What's happened?" It was Mr. Murdock's smooth voice. He and Mrs. Murdock had threaded their way through the crowd to reach them.

"My wife"—Dad turned to the Murdocks—"Ephraim says she's gone—left the trailer and taken the kids with her. I can't understand it! I know she was upset, but—"

Mrs. Murdock frowned. "She's moved the injured child? But why?—I told her not to do that!"

"Maybe she's here in the crowd," said Mr. Murdock. As if they had been dispatched, all four of them whirled in different directions, scanning the people who were milling around. But Mom and the children were nowhere to be seen.

"I know she's not in the store," said Mrs. Murdock tensely. "I've been in there, and I would have seen her."

"Let's go back to the trailer," said Dad. "Maybe she's come back."

"She might have heard all the yelling and gone to see what happened," said Mr. Murdock as they all started toward the trailer. "Did you know"—this to Ephraim—"that they've announced over the police radio that the accident—whatever it was—at Isla Conejo is under control? The Highway Patrol officer says we can all go home now."

The trailer was still empty. There was water on the floor —probably melted ice—and the butane light was still on, the screen door standing open. Mom's purse, as Ephraim had noted earlier, lay, unzipped, on the couch and beside it a jumble of things—lip salve, cough drops, keys, tissues, even her billfold with a couple of ten dollar bills crumpled into the money compartment.

Dad stared bleakly at the purse. "Carrie takes care of her things. She had to be running—and scared to death," he said, "to leave her purse like that."

Ephraim stared around at the trailer. "Dad? How'd she do it? She'd have to carry Jocelyn. And there's Caleb—"

"Bones must have carried Caleb. Or led him—" Dad turned and made a quick, thorough search of the trailer. He too found Kalijah and Turk in the little bathroom and quickly shut the door on them. "Come on—we've got to find them! I don't know why she's done this, but then this day would have driven anybody crazy!" He grabbed a flashlight from the counter by the sink and headed for the door, but as he jumped down the steps, followed by Ephraim, who slammed the trailer door, he plowed into the Murdocks.

"Did you find her?" asked Mrs. Murdock. She was out of breath and leaning heavily on her husband's arm.

"No," said Dad. "No sign of them. God, I don't know— And things scare her so bad—"

"But Helene told her to wait because we had a way to get the little girl to the hospital," said Mr. Murdock. "She could have been on her way right now if—"

Dad grabbed Ephraim by the shoulder again. Ephraim winced. The whole world was running itself by means of his left shoulder. He wondered how long it would hold out. "Ephraim," said his father, "get the other flashlight out of the pickup. We've got to look for her. She can't have gone very far."

"Maybe someone has seen her," suggested Mrs. Murdock, looking around. "Let's ask."

Suddenly Ephraim remembered the men in the black van. "There's nobody besides you parked right here near us except the guys who were in the black van," he said as he fished his father's big flashlight out of the glove compartment in the pickup. "And the van is still there, but the guys are gone."

Dad turned slowly, staring at Ephraim. "Didn't you say that one of the guys was handcuffed? To the tree?"

Ephraim straightened up suddenly. "Yeah. I did."

There was a short, chill silence. All of them were thinking, fast, about why a man would be handcuffed and what he might take it into his head to do when he was set free.

Ward Murdock spoke, hesitatingly. "I heard . . . Boone—the fellow who owns this place—say . . . something about . . . turning some guys who had sobered up . . . turning them loose. To go home . . ."

Dad stared past the trailer, past the edge of the parking lot and into the dark forest that met their eyes like a solid black wall.

"Then—" he said slowly, "I think I know where to look."

CHAPTER 20

The Murdocks followed them to the edge of the clearing. Beyond this point the land sloped up steeply and the trees and brush grew thick. Mrs. Murdock stopped and peered after them. "We'll get help!" she called.

As the Murdocks turned back, Dad halted, a few feet ahead of Ephraim. "There seems to be a kind of trail here," he said, flashing his light around.

"Are you sure she went this way?" asked Ephraim, coming up to stand by his father. The trail, as Dad called it, was hardly more than a gap in the manzanita brush that zigzagged up the side of the mountain.

"Can't tell, but it seems likely, since none of us saw her go toward the store or the road. And Carrie would look for a place like this to hide in if something . . . somebody . . . scared her. Let's go up here a ways. Keep your eyes open."

Dad, followed by Ephraim, lunged up the steep slope, his boots slithering on the slick dry grass.

"Can you see their tracks?" asked Ephraim, staring at the rough ground underfoot as he climbed after his father. The flashlight he held showed a scattered carpet of dry grass, pine needles and fallen manzanita leaves. He couldn't see anything that looked like the print of Mom's sandal or Bones' canvas shoes.

"Hard to tell. But I'm not good at that kind of thing, so we'll have to look hard. Keep watching."

"Dad—call her."

"Why didn't *I* think of that?" grunted his father in disgust, and then he sucked in a deep breath. "Carrie! Carrie! Carrie! Come back—everything's all right! Carrie! Carrie!"

"Mom!" yelled Ephraim. "Answer me! Where are you? Mom?"

They must have covered two or three hundred yards, zigzagging upward through heavy forest as the lights and sounds of the people at the store dropped farther and farther behind and below them, when Ephraim caught a flash of something over in a thicket just to the edge of his light. It was only a scrap of something, but it didn't belong there.

"Dad!" cried Ephraim. "Wait! I see something!"

Together Ephraim and his father raced to the thicket. Ephraim threw himself down and reached in under the dry, scratchy branches of a dead manzanita bush. His hand closed on something soft and white. He scrambled back to his feet and held the object out so his father could see it too.

"Bandage!" cried Ephraim triumphantly. "It's Bones' bandage! They did come this way!"

Dad poked the bandage with his finger. "Why did he lose it here?" he wondered.

"Probably ripped it off and threw it over here on purpose," said Ephraim. "He's leaving clues, like when we play cops and robbers at home. Or treasure hunt. He knows we'll be looking for them!"

As they started to turn back to the faint trail, Ephraim stuffed the bandage into his pocket. "Dad—" he began, but suddenly Dad's big hand fastened down again on his shoulder and his father hissed sharply for silence.

"Put your light out," whispered Dad.

Ephraim froze. Now he heard sounds—footsteps—

Before either of them could move, two dark figures burst through the screen of forest, moving fast, sliding and slithering down the steep slope not ten feet away from them. Neither of them spoke, though Ephraim and his father could hear the dry brush crackle and their grunts, as their feet struck and sank into the spongy mass underfoot. But even in the darkness, Ephraim knew who they were—or at least one of them.

"It's those two guys who were in the black van, Dad. I know it is!" whispered Ephraim.

The moment he had the words out, they saw, dimly, that one of the men had halted. He turned, and they thought he looked back up the mountain. "Listen, Ben," he said, "maybe—?"

"Maybe, hell," grunted the other man. "Let her go. No fun there—crazy screamin' dame, bunch of yellin' kids. Come on—we're goin' back to the city—have a *real* party there."

The first man hesitated for an aching moment, and then the speaker reached out and grabbed his arm. "Come on. We're going back and find Harley and Joan and Dick—have a good time there—come *on!*"

There was a small scuffle, and Ephraim and his father stood rooted like the thicket itself. The two men turned at last, and in a moment more they were out of earshot down the trail.

"Come on!" cried Dad. "Hurry! They're up here—someplace!"

A hundred yards farther on the trail petered out, but by that time they had found Caleb's bottle and a couple of Kleenex—pink, the kind Mom always carried in her pockets.

They were completely alone. If the Murdocks had mustered anyone to help in the search, no one appeared to be following them up the mountain yet. There were no sounds of tramping feet, no lights flashing. There were only the black forest, silent except for a very slight wind sighing through the tops of the pines, and the steep slope under their feet that threatened to trip them up with scattered stones, dead branches, patches of slick dry pine needles. Once again, even in the strangeness of night here on this black mountain, Ephraim had a fleeting thought that always before he had been with Mom and the little kids. Now he was here with Dad, a member of the rescue team. Bones, he thought suddenly—Bones has got to take my place—now that I'm out here with Dad.

"Dad," said Ephraim. "Call Mom again. You can yell louder than I can."

Dad took a deep breath. "Carrie! Carrie! Answer me! Carrie! Carrie!"

When he stopped calling, they both listened. Ephraim felt as if his ears were stretching out into bat wings on the sides of his head, he was trying so hard to hear something. He heard the wind and a sound that he knew was his own labored breathing, but nothing else.

"Up higher," said Dad. "Come on."

They started up the slope again, dodging between clumps of manzanita and big outcroppings of granite boulders whose flanks shone with silver speckles as the flashlights picked them out. Looking backwards, they could see the store, far below them, and people clustered here and there. And now, just beginning to move, a group of men heading toward the spot where they had entered the forest.

Dad grunted. "Here they come," he said. "Help. But if Carrie is as scared as I think she is, they'll only panic her more. Come on—we've got to find them."

They lunged up the slope again, levering themselves upward on tired legs, slashed across the face by branches, slipping, stumbling, gasping. By now they were climbing up a shallow cleft that seemed to direct their steps.

Ephraim had just decided to ask his father if they should separate so they could cover more ground, when Dad stopped to catch his breath.

And before Ephraim could speak, they heard it.

A little sound, only a tiny chink of sound falling into the great black void of the silence. And even though it was a sound he used to think he hated, Ephraim's heart leaped.

"Dad!" he cried sharply. "Listen! It's Caleb—crying!"

Ten, twenty yards up the slope, they halted. They knew they were very close. Caleb's crying was louder, closer, but didn't seem to come from any recognizable direction.

Ephraim groaned. "Dad—where *are* they? I can hear Caleb—"

"Carrie! Carrie!" cried Dad.

Suddenly Ephraim realized he must once more run ahead, take the lead. "Never mind Mom!" he cried sharply. "Call *Bones!*"

Dad grabbed a look at him. Then he nodded. "Bones!" he bellowed. "Answer me! Where are you? *Bones!*"

There was a long breath of silence. Then "Dad? Daddy? We're here—down in a hole!"

They leaped forward. Their lights ran crazily here and

there, up and down, till suddenly Dad let out a cry. "Jesus Christ!" He sprang forward half a dozen yards to where a huge pine tree—blasted by lightning maybe—had been toppled over. The great sprangly octopus of its roots, wrenched from their home in the soil, had left an enormous, gaping hole, half hidden by leaning bushes and long grass. Dad leaped to the edge of the hole and sank down with Ephraim at his side. They stabbed their lights into the ragged pit.

Mom sat huddled, Jocelyn in her arms, against the wall of roots. Bones crouched beside her, with Caleb in his arms.

Mom looked up. "They—chased us. They—grabbed at us —tore my clothes. I had—to run. Look—for a safe place— to hide!"

CHAPTER 21

The rescuers had reached them. There were enough men so Jocelyn and Caleb could be carried by two of them, and two more took Ephraim and Bones each by the arm to support and guide them. Accepting the help made Ephraim feel like a baby. But by now his legs were trembling from the climb, and he would rather be helped than fall headlong down the slope and make a real fool of himself. Dad and another man—Ephraim realized now that it was Mr. Murdock—were on each side of Mom, steadying her and talking quietly, although she seemed not to hear anything they said.

Dad explained to Mr. Murdock why Mom had bolted. "Boone must have sent somebody over to lock the guy up to keep him out of trouble until he sobered up. Then I guess he turned them loose, expecting them to head for home. Instead, they barged in on Carrie at the trailer. Tried to grab her, she says. So she took the kids and started to run. The guys got between her and the store, so she slipped away in the darkness and headed for the trees. And just kept on going."

"Could have killed herself. Or the kids," said Mr. Murdock tersely.

Dad said nothing.

After a moment, Mr. Murdock said, "Of course, people don't use their heads in a panic. Come to think of it—" He broke off for a moment while he stared down at the store and the parking lot and the crazy jumble of people and vehicles, the highway still clotted with traffic that now surged about equally in both directions— "Come to think of it—we've *all* been running around in a panic."

"Tell me about it," grunted Dad. "My God—will we *ever* get this mess straightened out—families home safe, traffic cleared, with nobody killed?"

"Time," said Murdock. "It will take time."

"How much time have we got?" asked Dad. "Look at us —my little girl hurt bad—one boy banged up and scared— my wife nearly out of her mind with worry. And I damn near shot a guy earlier tonight because he was going to steal my pickup."

Mr. Murdock gave Dad a startled look, then grunted as he half tripped over a rock. "Time," he repeated. "To find out what happened at Isla Conejo today. Time to take one hell of a long hard look at that place—see who is in charge there, and if they know what they're doing. Time to ask ourselves how much electricity we need—to run TVs, light up the Las Vegas strip, or pump water to grow lawns so we can mow them and throw the grass away. Maybe there are some new answers. There might even be some new questions. But all of this is going to take time, and we've got to work together."

"That's what Carrie tried to get me to do—" said Dad glumly. "Take time to find out what was going on. And I wouldn't do it."

"Well," said Mr. Murdock as he bent to hold back the stiff dry branches of a dead manzanita that leaned across their way, "I've never seen people come to grips with a problem yet until they got hurt by it."

At the bottom of the slope, they all halted for a moment. Dad turned to the men who had left their own urgent affairs behind and had climbed the steep dangerous mountain in the dark to help him and his family. "Thank you," he said gratefully, reaching out to shake each man's hand in turn, "thank you. We'll be all right now."

For a few more minutes they stood, Dad and Ephraim and the others, weary and out of breath, in a silent cluster near the trailer. Then, out of the dimness there appeared a lurching figure and Ephraim realized it was Mrs. Murdock.

"Ward?" said Mrs. Murdock. "Is that you? Is everybody all right?"

"Helene? Be careful, don't fall," said Murdock. "You're none too steady on your feet."

"Never mind about me," said Helene Murdock crisply. "Floyd's here—just landed. See? Over there—"

Automatically they all turned to see what Mrs. Murdock was pointing at. Then Ephraim gasped, and Bones, beside him, let out a yell. "Helicopter!" shouted Bones.

There on the cleared space beyond the row of little houses, highlighted by car lights and every other kind of light that could be mustered, stood a small, sleek, glistening black helicopter. Its blades were motionless, but Ephraim had a feeling that it had the power to just spring up from the earth and fly away, like a giant dragonfly on a summer day. So *this* was Mrs. Murdock's rabbit!

"We had planned to meet the helicopter here," said Mr. Murdock, "and I was going to send Helene to her sister's home in Reno till the emergency was over. I had told her I would wait for her here, but actually I was planning to go back and volunteer my services as a communications expert, if this whole thing developed into a prolonged emergency. That's my field—communications—and the helicopter belongs to my company. We use it for business trips. But Helene is so fond of kids—has taken care of them all her life. She had decided, even before they told us the accident at Isla Conejo was under control, that she was going to send the little girl and her mother to the hospital in Reno instead. The copter is very small—wouldn't carry all three of them together with the pilot. In any case, Helene and I can go back home now, so . . ."

Murdock turned, as if to give Mrs. Murdock the floor. The silver-haired woman moved toward Mom, and her voice was very gentle. "Mrs. Dearborn," she said, "you've been through so much today. It's cruel to press you, but you must pull yourself together. Your little girl is badly hurt—and the helicopter is the best way to get her to the hospital quickly. The nearest one is a long way off, and the roads—well—it's best to get the child there as soon as possible."

"Helicopter?" It was the first word Mom had said since they found her. "My baby—my Jocelyn—helicopter?"

"And you, of course," added Mrs. Murdock quickly, "will go with her."

There was a full breath of silence. All of them—Ephraim too—realized that Mom had almost reached her limit. After the long hours of fear and anxiety over the danger of a radiation leak from Isla Conejo, there was the attack by Dave beside the road, and now this last assault—her arms and neck showed livid welts where the man from the black van had raked her with his powerful hands. On top of all that, Mom was terrified of heights and flying. Was it possible that she could now find strength to climb into that black dragonfly over there and go away with Jocelyn over the dark mountains and canyons to a strange city and strange people?

Ephraim waited, afraid to breathe. All day he had been pushed forward again and again into things he couldn't do, but he did them. Now it was happening to Mom. Would she, *could* she, go forward now? For the first time in his life he felt *sorry* for her.

"Gabe?" Mom's voice was shaky.

"You—can do it, Carrie," said Dad. "You can do it."

"Floyd, the pilot, will radio ahead for an ambulance," said Mrs. Murdock. "You'll go straight to the hospital as soon as you land."

"And I'll follow with the boys," said Dad. "We'll catch up with you sometime tomorrow."

Mom drew in a shaky breath. "All right. I'll—I'll make it. I've got to be there to help Jocelyn. Come on—let's get started."

The wonderful rackety rhythmic clatter of the helicopter motor had faded away and its lights winked out as Floyd— a bearded giant in jeans and a red T-shirt with the sleeves torn off—guided it high over the first ridge and out onto the vast black sea of night sky.

Bones, Ephraim and Dad had stood in silence, watching and listening till it was gone. Then they made their way back from the makeshift landing pad and crossed the parking lot to Dad's pickup.

The Murdocks too had walked quietly, hand in hand, back to their Winnebago. They seemed to be making some plans, for presently they came over, and Ward Murdock handed Dad a crumpled slip of paper. "Our phone number and address in Lodi," he said. "We want to know how the little girl gets along."

Dad took the slip and folded it carefully. He put it into his billfold awkwardly, because he was holding Caleb and the baby was slowly relaxing into a limp sleepy bundle. "I'll never forget this," he said soberly to the older man and woman. "I don't know how on earth to thank you. Without your help, we might—we might have lost—" He choked into silence.

Mrs. Murdock reached out and patted Caleb's curly head on Dad's shoulder. "It's just a case of keeping your priorities straight and then using whatever you have at hand when something needs to be done," she said calmly. "By the way" —she straightened up and looked around at Ephraim and Bones—"speaking of getting things done, I hope you took good care of my fish."

Bones, who seemed to have lost most of his voice these last few hours, mumbled something. Mrs. Murdock shook her head, and Bones repeated, louder, "In the toilet."

"What!" cried Mrs. Murdock. "You flushed my Horatio down the toilet? How *could* you!"

Dad turned on Bones. "What the hell did you do that for?" he shouted.

"I didn't!" cried Bones. "He's—oh—come and look!"

Bones turned and yanked the trailer door open and scrambled up the steps. "Come *on,*" he urged the angry woman. "I'll show you!"

Mrs. Murdock climbed slowly up the steps after Bones, followed by Mr. Murdock and the others. They made quite a crowd inside the small trailer, and Bones had to maneuver carefully to open the narrow door into the little bathroom. As he peeled the door back, Kalijah and Turk poured out through the opening.

"If that cat ate my Horatio—" cried Mrs. Murdock. Bones grabbed Kalijah and thrust him into Ephraim's arms.

"Dad, your flashlight?" prodded Bones, and Dad clicked on his flash and shone it into the tiny bathroom. Bones stepped forward, raised the seat of the small yellow plastic toilet, reached in, and lifted out the glass bowl. Horatio was spinning around in his little home, safe and well, and gleaming like a piece of the sun.

Glowering a little, Bones turned to hand the bowl to Mrs. Murdock.

"I put the lid down," explained Bones, "so Kalijah couldn't get at him. There's no water standing in this kind of toilet, so the bowl didn't even get wet. It was the only place I could think of—Mom was yelling that we had to run, and I didn't have much time."

Dad sagged against the door of the bathroom. "Bones—I'm sorry. You used your head."

Behind them there was a muffled, squeaking sound, and when they turned, they found it was Mr. Murdock. He was laughing.

After a moment Mrs. Murdock began to smile a little, then a few chuckles broke through, and finally she laughed till her eyes watered, and she had to blow her nose. Dad laughed last, his cheek resting on Caleb's head.

Bones did not laugh. And after thinking about it, Ephraim didn't either. While the three grown-ups laughed out some of the fear and anger that danger had built up in them, Ephraim and Bones just stood looking at each other.

Ephraim knew he would never again see Bones the way he had only yesterday. It was Bones, he knew, who had carried or pulled Caleb up the mountain. Bones had kept up with Mom. And it was Bones who had had the presence of mind in the face of an attack to save Mrs. Murdock's little fish.

"Okay," grunted Ephraim grudgingly as the grown-ups' laughter echoed over their heads, "you . . . did okay. Now, if you would just quit hogging the TV . . ."

118

CHAPTER 22

They had packed the four of them—Dad, Ephraim, Bones and Caleb—into the front seat of the pickup. As they watched the rear lights of the Winnebago disappear to the west down the highway toward the valley, Ephraim shifted the blankets spread over his and Bones' legs. Caleb was asleep, and they would take turns holding him as Dad drove on eastward through the night, winding higher and higher through the mountain passes toward Reno and the hospital where they would find Jocelyn and Mom.

They had settled Turk and Kalijah and Corkey again in the trailer, locked the door, and as Dad started the pickup's engine, Bones peered at his wound. Ephraim snapped the lock down on the right-hand door. "Let's go, Dad," said Ephraim.

Caleb lay for now in Ephraim's lap, with his head, round as a soccer ball padded with golden curls, pillowed on an old coat Ephraim had stuffed down against the door. They had wrestled the baby into a dry diaper, stopping their ears against his crying. They could see that he was cruelly chafed from being wet so long. Dad had put so much baby powder on him to ease the pain that now the powder was sifting out everywhere, down his blue corduroy pants, out under his little shirt printed with hobby horses, and sprinkling down over the legs of Ephraim's jeans. The fragrance of the powder mingled oddly with the scent of pine coming in through the window, open an inch or two, the cigar Dad had lit, and the engine smells blowing back into the cab.

Ephraim thought, staring at their reflection in the side window, that this was the first time he had ever really held Caleb for any amount of time. What should have been familiar was truly strange—the size and shape of the baby's body, his little hands with fingers spread wide in sleep, the rapid rise and fall of his chest, the roundness of his butt and the little prods of his heels when he stirred and kicked.

Thinking these things reminded him of how strange Bones had felt when Ephraim reached out to comfort him, and the frightening limpness, the emptiness, of Jocelyn as she lay barely conscious.

Mom had seemed different this afternoon too—no, that wasn't quite right. Mom was . . . he struggled to find a way to say it to himself . . . Mom had been her same self, but more so. Nervous, anxious, easily frightened, Mom had been the same as always, but under these terrible pressures, she had revealed just how brittle she was, had made them understand once and for all that the limits of her strength and control would always come too soon. No matter what happened from now on, Ephraim knew, staring up the highway as the pickup lights made a capsule of brightness that slid endlessly ahead of them like a bead on a long string, no matter what happened now, none of them would ever see each other in quite the same way again.

"I should have told her *yes,*" Ephraim said suddenly, aloud.

Dad started. "Who? What?"

"Mrs. Murdock," said Ephraim. "She asked me if we would find a way to keep the world from blowing up. I should have said, *yes.*"

His father stared down the road. Bones had gone to sleep, his cheek resting against Dad's shoulder, and with the baby asleep, there were just the two of them awake and watching as the night flowed past.

"Well," said Dad slowly, "I think we'll make it—"

Ephraim turned to look at his father, and now saw that something else had changed too. Where always before Ephraim had been inside the fence with Bones and Jocelyn and Caleb—safe, but a prisoner—now he was outside the fence. The little ones would be there inside the comfortable old yard for some time yet, and he would be there to help take care of them part of the time, but now he could see that the yard fence no longer held him in. More and more, as time passed, he would be outside the fence, out here with Dad and people like the Murdocks—and people like Dave and the men in the black van? For a moment he was scared again, but the moment passed.

"Or I could have told her," said Ephraim, "I could have told her that the world is still here—but none of us will ever be the same."

And then he was silent, staring ahead into the darkness of the deep night.